# SolidWorks Workbook 2022 (Colored)

By
**Gaurav Verma**
**Matt Weber**
**(CADCAMCAE Works)**

Edited by
**Kristen**

ISBN # 978-1-77459-064-5

NOTICE TO THE READER

## DEDICATION

To teachers, who make it possible to disseminate knowledge
to enlighten the young and curious minds
of our future generations

To students, who are the future of the world

## THANKS

To my friends and colleagues

To my family for their love and support

# Table of Contents

## Chapter 4
## Surface Modeling Practical and Practice

## Chapter 5
## Sheetmetal Design Practical and Practice

## Chapter 6
## Assembly Design Practical and Practice

## Chapter 7
## Drafting Practical and Practice

# Preface

The **SolidWorks Workbook 2022**, is 1st edition of the book to be used as supplement practice book for SolidWorks Basic Learners. The book covers practical and practice questions. The methodology followed in this book for creating tutorials starts from beginner level and takes you to industrial level of designing models. The questions discussed in this book cover Sketching, Solid Modeling, Surface Modeling, Sheet Metal Design, Assembly, and Drafting. Some of the salient features of this book are:

## Instruction through illustration

The instructions to perform any action are provided by maximum number of illustrations so that the user can perform the actions discussed in the book easily and effectively. There are approximately 500 illustrations that make the learning process effective.

## Tutorial point of view

The book explains the concepts through the tutorial to make the understanding of users firm and long lasting. Each chapter of the book has tutorials that are real world projects.

## For Faculty

If you are a faculty member, then you can ask for video tutorials on any of the topic, exercise, tutorial, or concept. As faculty, you can register on our website to get electronic desk copies of our latest books, self-assessment, and solution of practical/ practice questions. Faculty resources are available in the **Faculty Member** page of our website (**www.cadcamcaeworks.com**) once you login. Note that faculty registration approval is manual and it may take up to two days for approval before you can access the faculty website.

## Formatting Conventions Used in the Text

All the key terms like name of button, tool, drop-down etc. are kept bold.

## Free Resources

Link to the resources used in this book are provided to the users via email. To get the resources, mail us at ***cadcamcaeworks@gmail.com*** or ***info@cadcamcaeworks.com*** with your contact information. With your contact record with us, you will be provided latest updates and informations regarding various technologies. The format to write us e-mail for resources is as follows:

Subject of E-mail as ***Application for resources of ............... Black Book***.
You can give your information below to get updates on the book.
***Name:***
***Course pursuing/Profession:***
***Contact Address:***
***E-mail ID:***

Note:
**SolidWorks 2022 Black Book** can be used as **SolidWorks CAD companion** with this book for learning about modeling tools.

## About Author

The author of this book, Matt Weber, has authored many books on CAD/CAM/CAE books. He has authored **SolidWorks 2022 Black Book** as a CAD companion for this book. **SolidWorks Simulation 2022 Black Book** covers details of simulation and the SolidWorks 2022 Black Book covers all the tools and techniques of modeling. The author has hand on experience on almost all the CAD/CAM/CAE packages. If you have any query/doubt in any CAD/CAM/CAE package, then you can directly contact the author by writing at cadcamcaeworks@gmail.com

The technical editor of the book, Gaurav Verma, has authored books on different CAD/CAM/CAE packages. He has authored SolidWorks Flow Simulation 2022 Black Book, SolidWorks Electrical 2022 Black Book,  Creo Manufacturing 4.0 Black Book, MasterCAM 2022 for SolidWorks Black Book, AutoCAD Electrical 2022 Black Book, Autodesk Inventor 2022 Black Book, Autodesk Fusion 360 Black Book, and many others.

**For Any query or suggestion**

If you have any query or suggestion please let us know by mailing us on *cadcamcaeworks@gmail.com* or *info@cadcamcaeworks.com*. Your valuable constructive suggestions will be incorporated in our books and your name will be addressed in special thanks area of our books.

# Chapter 1

# Introduction

## Topics Covered

The major topics covered in this chapter are:

- *Installing SolidWorks.*
- *Starting SolidWorks.*
- *Starting a new document.*
- *Terminology used in SolidWorks.*
- *Opening a document.*
- *Closing documents.*
- *Basic Settings for SolidWorks*
- *Workflow in Industries using the SolidWorks*

## INSTALLING SOLIDWORKS

You can get SolidWorks software installation files in two ways, downloading from website or getting DVD of software from your reseller.

* If you are installing SolidWorks using the CD/DVD provided by Dassault Systemes then go to the folder containing **setup.exe** file and right-click on **setup.exe** in the folder. A shortcut menu is displayed on the screen; refer to Figure-1.

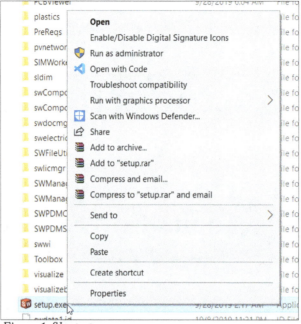

*Figure-1. Shortcut menu*

* Select the **Run as Administrator** option from the menu displayed; refer to Figure-1.
* Select the **Yes** button from the dialog box displayed. The **SolidWorks Installation Manager** will be displayed. Follow the instructions given in the dialog box. Note that you must have the **Serial Number** with you to install the application. To get more about installation, double-click on the **Read Me** documentation file in the Setup folder.
* If you have downloaded the software from Internet, then you are required to browse in the **SolidWorks Download** folder in the **Documents** folder of Computer. Open the folder of latest version of software and then run **setup.exe**. Rest of the procedure is same.

## STARTING SOLIDWORKS

* To start SolidWorks in Windows 10 from **Start** menu, click on the **Start** button in the Taskbar at the bottom left corner, click on the **SolidWorks** folder. In this folder, select the SolidWorks icon; refer to Figure-2.

*Figure-2. Start menu*

- While installing the software, if you have selected the check box to create a desktop icon then you can double-click on that icon to run the software.
- If you have not selected the check box to create the desktop icon but want to create the icon on desktop, then drag and drop the **SolidWorks** icon from the Start menu on the desktop.

After you perform the above steps, the SolidWorks application window will be displayed; refer to Figure-3.

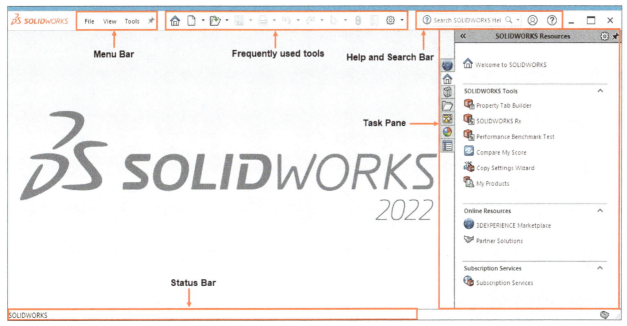

*Figure-3. SolidWorks application window*

## STARTING A NEW DOCUMENT

There are various methods to start a new document in SolidWorks. The procedure to start a new document in SolidWorks is given next.

1. Click on the **New** button in the **Menu Bar** 🗋 .
Or
2. Move the cursor on the left arrow near the **SolidWorks icon**; refer to Figure-4 and then click on the **File** menu button. The **File** menu will be displayed, click on the **New** button; refer to Figure-4.
Or
3. Press **CTRL** and **N** together from the Keyboard.

*Figure-4. File menu*

- After performing any of the above steps, the **New SOLIDWORKS Document** dialog box will be displayed as shown in Figure-5.

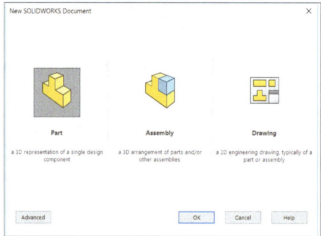

*Figure-5. New SOLIDWORKS document*

- If you are creating first document after installing SolidWorks then **Units and Dimension Standard** dialog box will be displayed; refer to Figure-6.

*Figure-6. Units and Dimension Standard dialog box*

- Select the unit system and dimension standards that you want to use while creating documents in SolidWorks. These options will be set as default for later documents. You can change these parameters later using **System Options**. Click on the **OK** button from the dialog box.

There are three buttons available in this dialog box; **Part**, **Assembly**, and **Drawing**.

The **Part** button is used to create Solid, Surface, Sheetmetal, Mesh and other types of models.

The **Assembly** button is used to create Assemblies.

The **Drawing** button is used to create drawings from the part models or assemblies.

You will learn more about solids, surfaces, assemblies, and drawings later in the book.

Note that the building blocks of CAD are solid models. In SolidWorks, solid models are created by using the tools available in the **Part** mode. You can start with the **Part** mode by selecting the **Part** button in the **New SOLIDWORKS Document** dialog box.

- Double-click on the **Part** button to start the part modeling environment of SolidWorks. On doing so, the application interface will be displayed as shown in Figure-7.

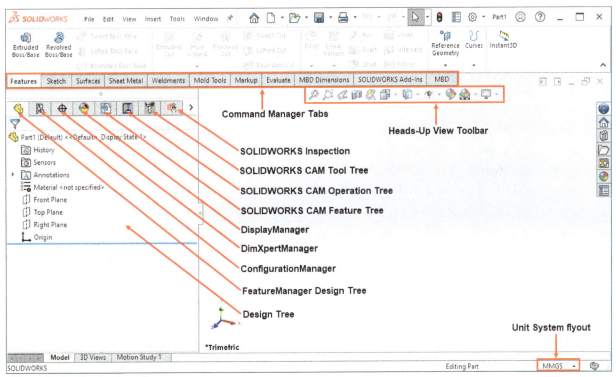

*Figure-7. Application interface*

The tools available in the **Part, Assembly**, and **Drawing** mode are compiled in the form of **CommandManagers**. To display or hide a **CommandManager** from **Ribbon**, right-click on a **CommandManager** in the **Ribbon**. A shortcut menu will be displayed. Hover the cursor over **Tabs** option, the list of **CommandManagers** will be displayed as shown in Figure-8. Select the **CommandManager** that you want to add to or remove from the **Ribbon**. Note that a tick mark is displayed before name of **CommandManager** in the list if the **CommandManager** is there in the **Ribbon**. Various **CommandManagers** available in SolidWorks will be discussed next.

*Figure-8. Shortcut menu for CommandManager*

## Part Mode CommandManagers

A number of **CommandManagers** can be invoked in the **Part** mode. These **CommandManagers** with their functioning are discussed next.

### Sketch CommandManager

The tools available in this **CommandManager** are used to draw sketches for creating solid/surface models. This **CommandManager** is also used to add relations and smart dimensions to the sketched entities. The **Sketch CommandManager** is shown in Figure-9.

*Figure-9. Sketch CommandManager*

### Features CommandManager

This **CommandManager** provides all modeling tools that are used for feature-based solid modeling. The **Features CommandManager** is shown in Figure-10.

*Figure-10. Features CommandManager*

## MBD Dimensions CommandManager

This **CommandManager** is used to add dimensions and tolerances to the features of a part. The **MBD Dimensions CommandManager** is shown in Figure-11.

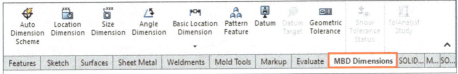

*Figure-11. MBD Dimensions CommandManager*

## Sheet Metal CommandManager

The tools in this **CommandManager** are used to create the sheet metal parts. The **Sheet Metal CommandManager** is shown in Figure-12. If this **CommandManager** is not added in the **Ribbon**, then right-click on any of the **CommandManager** tab and select the **Sheet Metal** option from the menu; refer to Figure-13.

*Figure-12. Sheet Metal CommandManager*

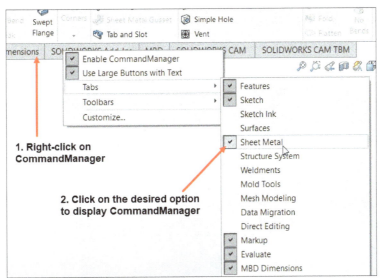

*Figure-13. Adding hidden tabs in Ribbon*

## Mold Tools CommandManager

The tools in this **CommandManager** are used to design a mold and split core & cavity steel. The **Mold Tools CommandManager** is shown in Figure-14.

*Figure-14. Mold Tools CommandManager*

## Evaluate CommandManager

This **CommandManager** is used to measure entities, perform analysis, and so on. The **Evaluate CommandManager** is shown in Figure-15.

*Figure-15. Evaluate CommandManager*

## Surfaces CommandManager

This **CommandManager** is used to create complicated surface features. The **Surfaces CommandManager** is shown in Figure-16.

*Figure-16. Surfaces CommandManager*

## Direct Editing CommandManager

This **CommandManager** consists of tools (Figure-17) that are used for editing a feature.

*Figure-17. Direct Editing CommandManager*

## Data Migration CommandManager

This **CommandManager** consist of tools (Figure-18) that are used to work with the models created in other packages or in different environments.

*Figure-18. Data Migration CommandManager*

## Weldments CommandManager

This **CommandManager** is used to create welding joints in the model and assembly. The **Weldments CommandManager** is shown in Figure-19.

*Figure-19. Weldments CommandManager*

## SOLIDWORKS MBD CommandManager

This **CommandManager** is used to apply Model Based Dimension which means the dimensions are directly applied to model while skipping the steps of generating drawings. Use of SolidWorks MBD in manufacturing industry requires electronic gadgets at shop floor for production and quality checks. These gadgets should be capable of displaying CAD eDrawings. The **CommandManager** is shown in Figure-20.

*Figure-20. SOLIDWORKS MBD CommandManager*

## Render Tools CommandManager

The tools in **Render Tools CommandManager** are used to render the image of current model using appearance parameters specified. The **CommandManager** is shown in Figure-21.

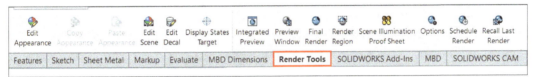

*Figure-21. Render Tools CommandManager*

## SOLIDWORKS CAM CommandManager

The tools in **SOLIDWORKS CAM CommandManager** are used to generate NC programs for CNC machines. The **CommandManager** is shown in Figure-22.

*Figure-22. SOLIDWORKS CAM CommandManager*

## SOLIDWORKS CAM TBM CommandManager

The tools in **SOLIDWORKS CAM TBM CommandManager** are used to automatically generate toolpaths and operation settings based on specified tolerances and dimension in the MBD; refer to Figure-23.

*Figure-23. SOLIDWORKS CAM TBM CommandManager*

## SOLIDWORKS Inspection CommandManager

The tools in **SOLIDWORKS Inspection CommandManager** are used to create inspection drawings and reports for various standards and production part approval process(PPAP); refer to Figure-24. You can also use SolidWorks Inspection for First Article Inspection (FAI) and in process inspection applications.

*Figure-24. SOLIDWORKS Inspection CommandManager*

## Sketch Ink CommandManager

The tools in **Sketch Ink CommandManager** are create sketch free-hand using either stylus or fingertips on touch screen devices; refer to Figure-25.

*Figure-25. Sketch Ink CommandManager*

## Markup CommandManager

The tools in **Markup CommandManager** are used to create free hand markings in the part, assembly, or drawing; refer to Figure-26.

*Figure-26. Markup CommandManager*

# Assembly Mode CommandManagers

The tools in **CommandManagers** of the **Assembly** mode are used to assemble the components. The **CommandManagers** in the **Assembly** mode are discussed next.

## Assembly CommandManager

This **CommandManager** is used to insert a component and apply various types of mates to the assembly. The **Assembly CommandManager** is shown in Figure-27.

*Figure-27. Assembly CommandManager*

## Layout CommandManager

The tools in this **CommandManager** (Figure-28) are used to create and edit blocks.

*Figure-28. Layout CommandManager*

## Drawing Mode CommandManagers

You can invoke a number of **CommandManagers** in the **Drawing** mode. The **CommandManagers** that are extensively used during the designing process in this mode are discussed next.

### Drawing CommandManager

This **CommandManager** is used to generate the drawing views of an existing model or an assembly. The **Drawing CommandManager** is shown in Figure-29.

*Figure-29. Drawing CommandManager*

### Annotation CommandManager

The **Annotation CommandManager** is used to generate the model items and to add notes, balloons, geometric tolerance, surface finish symbols, and so on to the drawing views. The **Annotation CommandManager** is shown in Figure-30.

*Figure-30. Annotation CommandManager*

The commands available in these **CommandManagers** will be discussed one by one later in this book.

## OPENING A DOCUMENT

Like creating new documents, there are many ways to open documents. Some of them are discussed next.

- Click on the **Open** button in the **Menu Bar** 📂 or Move the cursor on the **SolidWorks icon** and then click on the **File** > **Open** button from the menu or press **CTRL** and **O** together from the Keyboard.
- After performing any of the above step, the **Open** dialog box will be displayed; refer to Figure-31.
- Select the file type of your file from the **File Type** fly-out `Custom (*.prt;*.asm;*.drw;*.sld|` ▾ in the bottom right corner of the dialog box.
- Browse to the folder in which you have saved the file and then double-click on it to open.

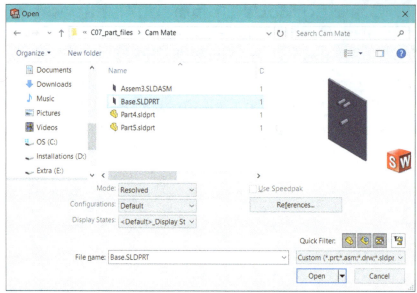

*Figure-31. Open dialog box*

Most of the time people close the application to close the current file and then restart the application to start working again. In SolidWorks, you can close the current file while the application remains open. The steps to do so are given next.

Note that there is no separate tool in SolidWorks to import models created in other software. You need to use **Open** tool to import the non native files.

## CLOSING A DOCUMENT

To close a document, there are two ways displayed in Figure-32.

- Open the **File** menu and then click on the **Close** option from it or click on the **Close** button ⊠ at the top-right of the current viewport (Viewport is the area in which the model is displayed.) or press CTRL+W.

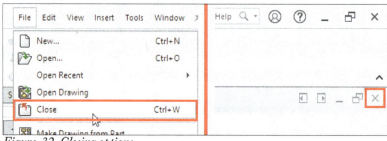

*Figure-32. Closing options*

- If you have done some editing in the document then a dialog box will be displayed prompting you to save the document; refer to Figure-33.

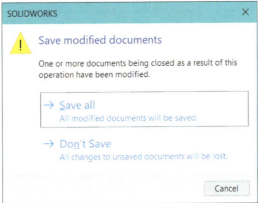

*Figure-33. Save prompt*

- Click on the **Save all** button to save the changes or click on the **Don't Save** button to reject the changes.

## SAVING THE DOCUMENTS

There are three tools available in **File** menu to save document in SolidWorks; **Save**, **Save As**, and **Save All**. If you are working on a model and click on the **Save** tool for the first time after starting the new document then the **Save** tool will work as **Save As** tool. The **Save As** tool allows you create a new copy of the document in different formats. The **Save All** tool is used to save all the modified documents currently open in SolidWorks. The procedure to use the **Save As** tool is discussed next.

- Click on the **Save As** tool from the **File** menu. The **Save As** dialog box will be displayed; refer to Figure-34.

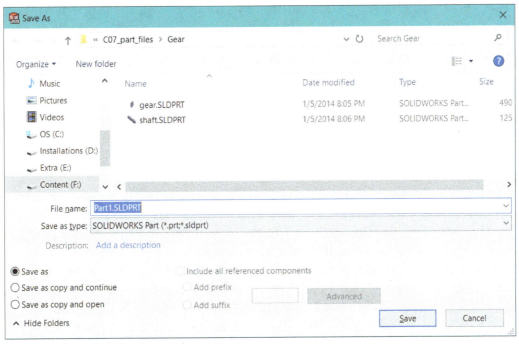

*Figure-34. Save As dialog box*

- Specify desired name for the file in the **File name** edit box.
- Click in the **Save as** type drop-down and select the desired format for file; refer to Figure-35.

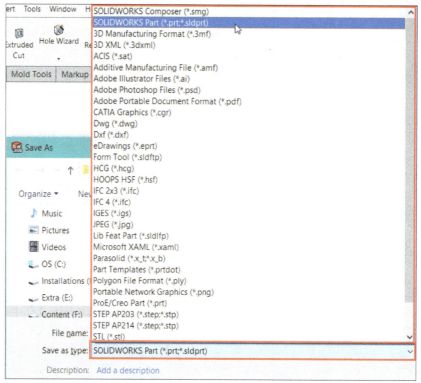

*Figure-35. Save as type drop-down*

- Select the **Save as** radio button to save the current open file with desired settings. Select the **Save as copy and continue** radio button to save a new copy of currently open file and do not open the new copy while leaving the original file unchanged. Select the **Save as copy and open** radio button to save a new copy of currently open file and open the new copy while leaving the original file unchanged.

- Click in the **Description** field and specify desired description for the file.

- If you are saving an assembly file then **Include all reference components** check box will be active. Select this check box to add a prefix/suffix to the assembly components names and define other parameters for saving assembly components files. The related options will become active. Select the desired radio button; **Add prefix** or **Add suffix** and then specify desired text in the next edit box. Click on the **Advanced** button. The **Save As with References** dialog box will be displayed; refer to Figure-36. Select the **Nested view** or **Flat view** radio button to check assembly components in nested view or flat view.

- Select the **Include broken references**, **Include Toolbox parts**, and/or **Include virtual components** check boxes to include respective parts while saving them.

- Click in the **Specify folder for selected items** edit box and specify the location where you want to save assembly components.

- Select the **Save all as copy (opened documents remain unaffected)** check box if you want to save new copy of all the assembly files in specified folder location.

- Double-click in desired field of table in the dialog box to change it.

- Click on the **Save All** button from the dialog box to save assembly files.

Note that once you have saved SolidWorks assembly or part file then next time clicking on **Save** tool or pressing **CTRL+S** will not display the **Save As** dialog box. It will directly save the files.

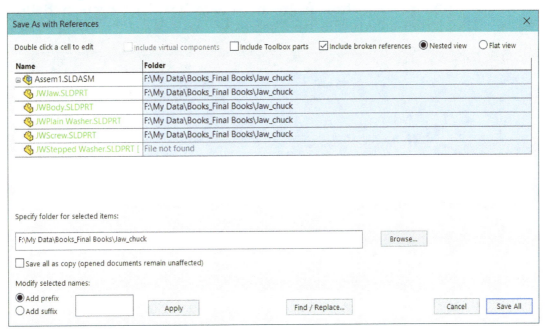

*Figure-36. Save As with References dialog box*

# PRINTING DOCUMENT

The **Print** tool in **File** menu is used to print documents on paper or in different formats. The procedure to use this tool is given next.

- Click on the **Print** tool from the **File** menu or press **CTRL+P**. The **Print** dialog box will be displayed; refer to Figure-37.

*Figure-37. Print dialog box*

- Select the desired printer from the **Name** drop-down in the **Document Printer** area of the dialog box.

- Click on the **Properties** button from the dialog box. The respected Properties dialog box will be displayed; refer to Figure-38.

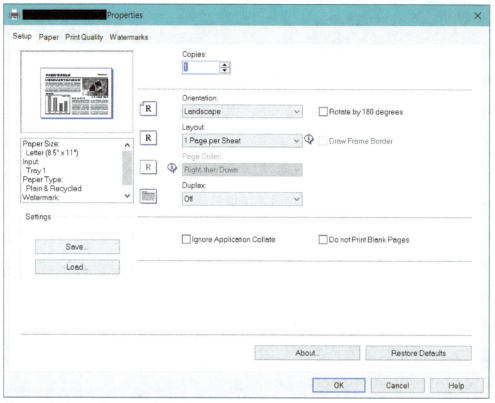

*Figure-38. Printer Properties dialog box*

- Set the other parameters as desired for printing like number of copies, paper size, orientation, resolution, and so on. Click on the **OK** button from the dialog box to apply properties.

## Applying Page Setup

- Click on the **Page Setup** button from the **Print** dialog box. The **Page Setup** dialog box will be displayed; refer to Figure-39.

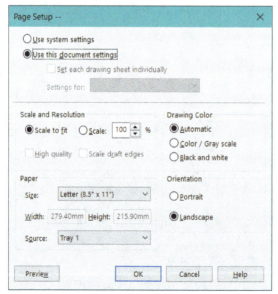

*Figure-39. Page Setup dialog box*

- Select the **Use system settings** radio button or the **Use this document settings** radio button to use respective page settings for printing.
- Set the other parameters as desired like print page source, paper size, scale, resolution, drawing color, and so on.
- Click on the **OK** button from the dialog box. The **Print** dialog box will be displayed again.

## Setting Header/Footer for Print Pages

- Click on the **Header/Footer** button from the **Document Options** area of the dialog box. The **Header/Footer** dialog box will be displayed; refer to Figure-40.

*Figure-40. Header/Footer dialog box*

- Select the desired options from the **Header** and **Footer** drop-downs. If you want to create a custom header/footer then click on the **Custom Header** button. The **Custom Header** dialog box will be displayed; refer to Figure-41.
- Click in the desired section and specify the custom text to be displayed in header or footer. You can also use the buttons like **Page Numbers**, **Date**, and so on to specify custom data. After setting desired parameters, click on the **OK** button. The **Print** dialog box will be displayed again.

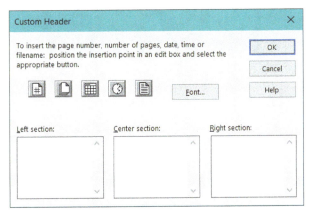

*Figure-41. Custom Header dialog box*

## Print Margins

- Click on the **Margins** button from the **System Options** area of the dialog box. The **Margins** dialog box will be displayed; refer to Figure-42.

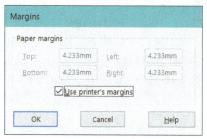

*Figure-42. Margins dialog box*

- Clear the **Use printer's margins** check box if you want to specify custom margins to keep object within paper boundary and specify desired parameters in the **Paper margins** area.
- Click on the **OK** button from the dialog box to apply parameters.

## Print Range and other Parameters

- Select the desired radio button from the **Print range** area and define respective parameters. If your document has multiple sheets for printing generally in case of printing drawings then **Current sheet** and **Sheets** radio button will be displayed.
- Specify the desired number for prints to be produced in the **Number of copies** edit box.
- Select the **Print background** check box from the dialog box if you want to print the background also.
- Select the **Print to file** check box if you want to save a printer file.
- Similarly, set the other parameters and click on the **OK** button from the **Print** dialog box.

## Publishing to eDrawings

The **Publish to eDrawings** tool is used to publish model in eDrawings Viewer. eDrawings is a software developed by Dassault Systemes Corporation for sharing models with management. The procedure to publish model is given next.

- Click on the **Publish to eDrawings** tool from the **File** menu. The **Save Configurations to eDrawings file** dialog box will be displayed; refer to Figure-43.

*Figure-43. Save Configurations to eDrawings file dialog box*

- Select the check boxes for configurations to be included in the eDrawing file.
- Click on the **Options** button from the dialog box. The **System Options** dialog box will be displayed; refer to Figure-44.

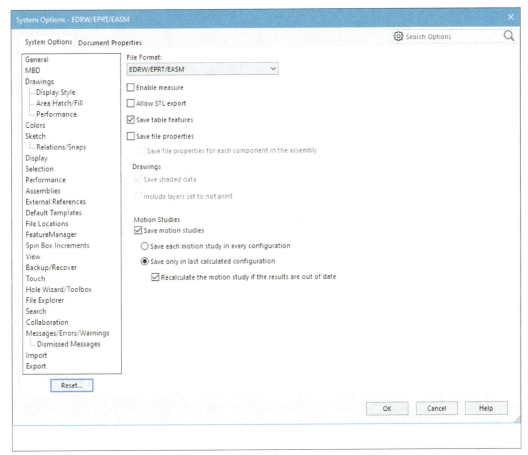

*Figure-44. System Options dialog box*

- Select the **Okay to measure this eDrawings file** check box if you want to allow measurement of geometry in eDrawing.
- Select the **Allow export to STL for parts & assemblies** check box to allow saving of eDrawing as STL from the eDrawing Viewer software.
- Similarly, set the other parameters as desired in the dialog box and click on the **OK** button.

## Specifying Password

- If you want to set password for opening eDrawing file then click on the **Password** button from the **Save Configurations to eDrawings file** dialog box. The **Password** dialog box will be displayed; refer to Figure-45.

*Figure-45. Password dialog box*

- Select the **Password Required to Open Document** check box. The edit box to define password will be activated.
- Specify the desired password and input the same in **Confirm Password** edit box. Click on the **OK** button from the dialog box.

After specifying desired parameters, click on the **OK** button from the dialog box. The model will open in eDrawing Viewer; refer to Figure-46. The options of eDrawings Viewer are out of scope of this book.

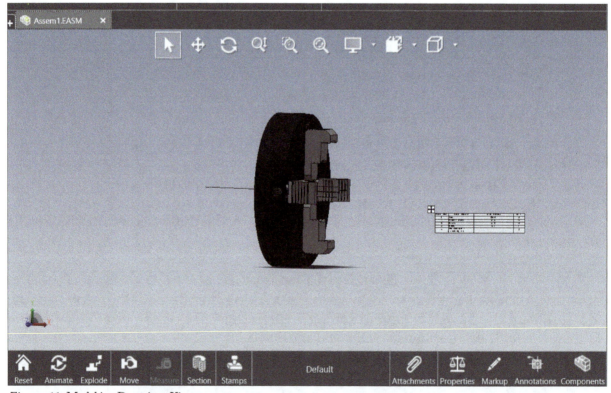

*Figure-46. Model in eDrawings Viewer*

## PACKING FILES FOR SHARING

The **Pack and Go** tool in **File** menu is used to package the files for sharing with colleagues or client. The procedure to use this tool is given next.

- Click on the **Pack and Go** tool from the **File** menu. The **Pack and Go** dialog box will be displayed; refer to Figure-47.

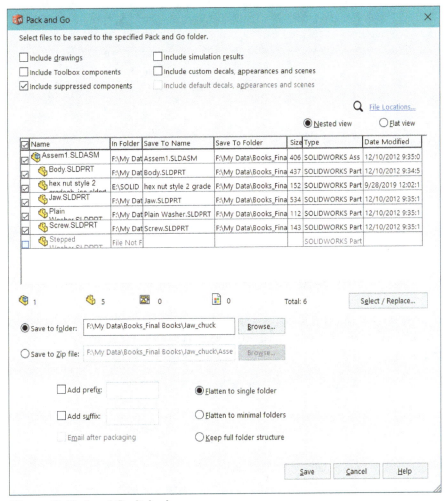

*Figure-47. Pack and Go dialog box*

- Select check boxes for documents to be included in the package.
- Select the **Save to folder** radio button if you want to create a new folder for documents or select the **Save to Zip file** radio button. Specify the desired location for file or folder in adjacent edit box.
- Select the **Add prefix** and **Add suffix** check boxes if you want to add prefixes and suffixes. After selecting the check boxes, specify desired text in adjacent edit boxes.
- Select the **Email after packaging** check box if you want to email your client after packaging files.
- Click on the **Save** button to save the file.

## RELOADING FILES

The **Reload** tool in **File** menu is used to reload the files that have been modified outside the SolidWorks software or the files that have not been updated automatically. The procedure to use this tool is given next.

- Click on the **Reload** tool from the **File** menu. The **Reload** dialog box will be displayed; refer to Figure-48.

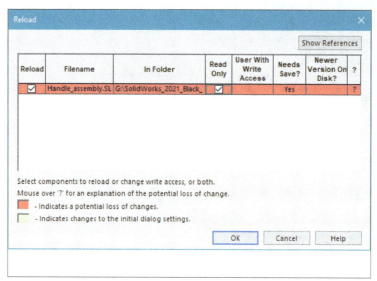

*Figure–48. Reload dialog box*

- Select the **Show full paths** check box to check full path of file to be reloaded.
- Click on the **OK** button from the dialog box. An information box will be displayed; refer to Figure-49.

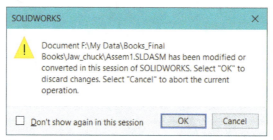

*Figure–49. Information box for reloading files*

- Click on the **OK** button from the dialog box.

## REPLACING COMPONENT IN ASSEMBLY

The **Replace** tool in **File** menu is used to replace selected component in the assembly. The procedure to use this tool is given next.

- Click on the **Replace** tool from the **File** menu. The **Replace PropertyManager** will be displayed; refer to Figure-50.
- Select the component to be replaced from **FeatureManager Design Tree**.
- If there are multiple instances of the same component and you want to replace all the instances then select the **All instances** check box.
- Click on the **Browse** button from the **With this one** area of **PropertyManager**. The **Open** dialog box will be displayed.
- Select the desired component by which original component will be replaced and click on the **Open** button.
- Select the desired radio button from the Options rollout. If you want to select configuration of new component with name matching to original one then select the **Match name** radio button. Select the **Manually select** radio button if you want to manually select configuration of component.

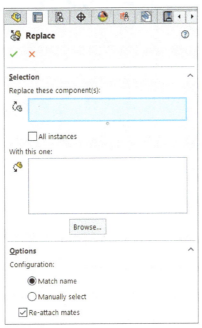

*Figure-50. Replace PropertyManager*

- Select the **Re-attach mates** check box if you want to reapply the mates. If you have selected this check box then you will be asked to select references for different mates.
- After setting desired parameters, click on the **OK** button from the **PropertyManager**. The **Mated Entities PropertyManager** will be displayed; refer to Figure-51.

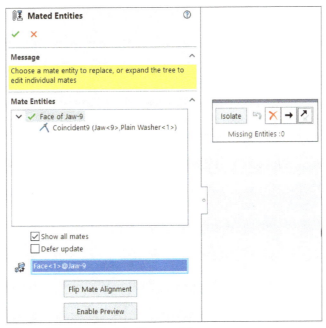

*Figure-51. Mate Entities PropertyManager*

- Set the references as desired and click on the **OK** button from **PropertyManager**. The mates will be applied. You will learn more about mates later in the book.

You can use the **Find References** tool from the **File** menu in the same way.

## CHECKING SUMMARY OF DOCUMENT

The **Properties** tool in **File** menu is used to check and manage summary information of file. The procedure to use this tool is given next.

*   Click on the **Properties** tool from the **File** menu. The **Summary Information** dialog box will be displayed; refer to Figure-52.

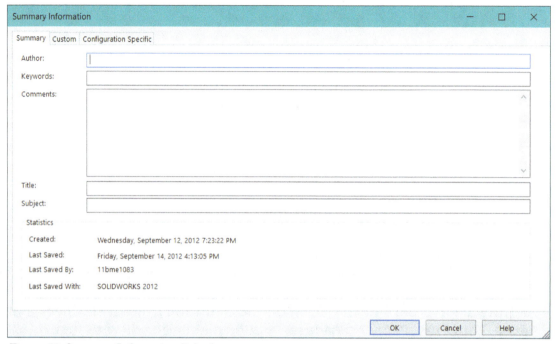

Figure-52. Summary Information dialog box

*   Specify the parameters like name of design author, title of design, comments and so on for the model.
*   Click on the **Custom** tab to define custom summary data.
*   Click on the **Configuration Specific** tab from the dialog box to define various bill of materials related parameters like material name, cost of material, and so on.
*   Click on the **OK** button after setting desired parameters.

## CUSTOMIZING MENUS

You can customize the menu by selecting the **Customize Menu** option from the **File** menu. The options to customize menu will be displayed; refer to Figure-53. Clear the check boxes for options to be removed from the menu. After setting desired options for menu, click in the empty drawing area. The menu will be updated accordingly.

To exit the software, click on the **Exit** tool from the **File** menu.

Till this point, you have learned the basic file handling operations and you have some idea about the interface of SolidWorks. Now, we will discuss about some basic settings that are required for easy working with SolidWorks.

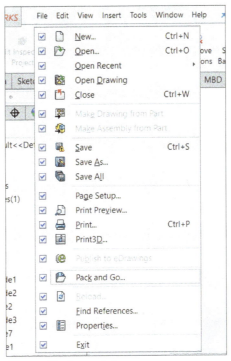

*Figure-53. Customizing menu*

# BASIC SETTINGS OF SOLIDWORKS

All the settings of SolidWorks are compiled in the **Options** dialog box. The steps to change the settings for SolidWorks are given next.

- Click on the **Options** option in the **Tools** menu or click on the **Options** button ⚙ from the **Menu Bar**.
- On performing the above step, the **System Options** dialog box will be displayed as shown in Figure-54. Note that if you have a document opened then the **Document Properties** tab is also added with the **System Options** tab. To get the detail about each and every option, you need to refer to SolidWorks Help Documentation. In this section, we will discuss about some of the important options that are generally required.
- Click on the **Sketch** option in the left of the dialog box, select the **Enable on screen numeric input on entity creation** check box from the right to enter dimensions while creating the sketch. Also, select the **Create dimension only when value is entered** check box to create dimensions only when you have manually entered the dimension value.
- If you want to use only fully defined sketches for creating features in SolidWorks then select the **Use Fully defined sketches** check box. Fully defined sketches are those sketch which have all their entities dimensioned or constrained.
- Select the **Auto-rotate view normal to sketch plane on sketch creation and sketch** edit check box to automatically make the sketching plane parallel to screen.
- Click on **Relations/Snaps** in the left of the dialog box and then select the **Enable snapping** check box to enable auto snapping to the key points.

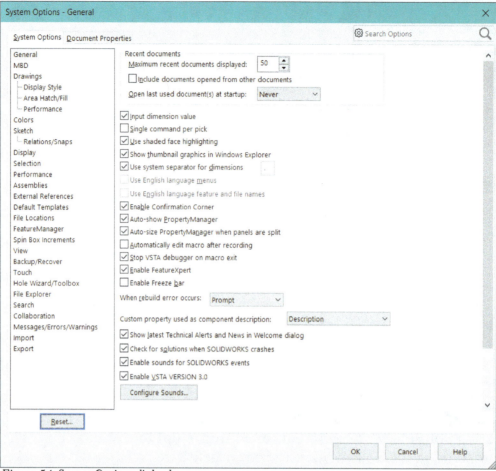

*Figure-54. System Options dialog box*

- Click on the **Document Properties** tab if you have any document opened in the viewport. Click on the **Units** option from the left. The **Options** dialog box will be displayed as shown in Figure-55.

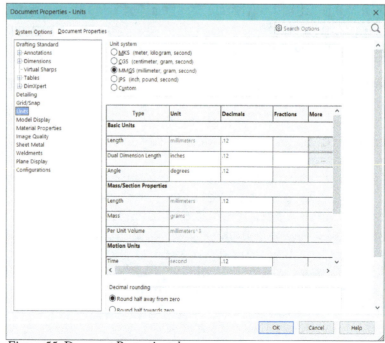

*Figure-55. Document Properties tab*

- Select the desired radio button from the right to set the unit system for the current document.
- Click on the **Drafting Standard** option from the left area of the dialog box and select the desired dimensioning standard from the **Overall drafting standard** drop-down; refer to Figure-56.

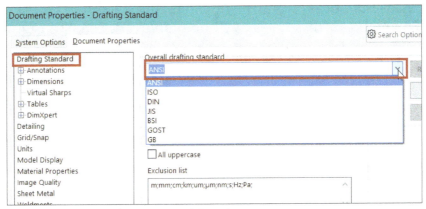

*Figure-56. Overall drafting standard drop-down*

- Similarly, you can set the other options in the dialog box. Note that we will be revisiting this dialog box many times in this book.
- Click on the **OK** button from the bottom of the dialog box to save the settings.

You can also change the units of document by selecting the desired option from the list displayed on clicking on the **Unit system** flyout at the bottom of the viewport; refer to Figure-57.

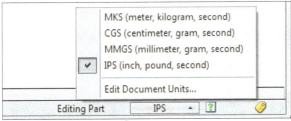

*Figure-57. Unit system flyout*

The other options of the **Options** dialog box are discussed in appendix pages.

## MOUSE BUTTON FUNCTION

### Rotate View (parts and assemblies only)
- To rotate the model view: Drag with the middle mouse button.
- To rotate about a vertex, edge, or face: Middle-click a vertex, edge, or face; then middle-drag the pointer.

### Pan
Hold down **CTRL** key and drag with the middle mouse button. (In an active drawing, you do not need to hold down **CTRL** key.)

### Zoom In/Out
Hold down **Shift** and drag the middle mouse button up/down or scroll the mouse wheel downward/upward to zoom in/out.

## LOADING ADD-INS

Add-Ins are used to allow external functions in SolidWorks. Like, you can use SolidWorks Electrical, SolidWorks Simulation, SolidWorks PCB, MasterCAM etc. The procedure to load Add-Ins is given next.

*   Click on the **Add-Ins** option from the **Options** drop-down in the **Menu Bar**; refer to Figure-58. The **Add-Ins** dialog box will be displayed; refer to Figure-59.

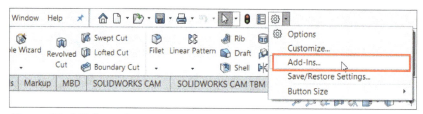

*Figure-58. Add-Ins option*

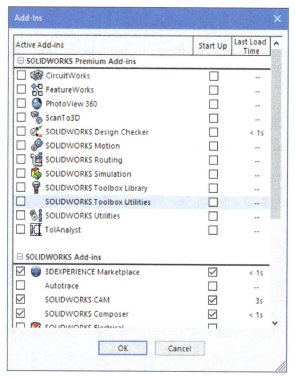

*Figure-59. Add-Ins box*

*   Select the left side check box for Add-In that you want to be loaded now. If you want to load an Add-In at startup of SolidWorks then select the right-side check box for it.
*   Click on the **OK** button from dialog box to apply the settings.

## SEARCH TOOLS

The **Search box** at the top-right corner of the application window is a multipurpose tool. You can use this search box to search for help content, commands, cad models, training files etc. The method to use **Search box** for searching commands is given next. You can apply the same method to search other things.

- Hover the cursor on **SOLIDWORKS** logo at the top-left corner of application window and click on the **Commands** option from the **Search** cascading menu of **Help** menu; refer to Figure-60.

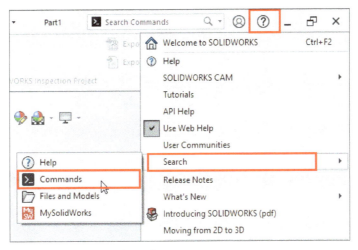

*Figure-60. Commands search option*

- Click in the **Search box** and type few characters of command that you are searching. A list of tools with typed characters will be displayed; refer to Figure-61.

*Figure-61. List of tools in search box*

- Click on the eye icon displayed next to tool in the list. The cursor will move to the tool location and an arrow will be displayed pointing to the tool; refer to Figure-62. If you click on the tool in the list then it will be activated directly.

*Figure-62. Location of tool displayed*

Note that you can pin the menu bar to display all the menus by clicking on the **Pin** button; refer to Figure-63.

*Figure-63. Pinned menubar*

## Login to SolidWorks

The **Login to SOLIDWORKS** button in the **Search Bar** is used to login to your SolidWorks account using the software; refer to Figure-64. Using this tool to login allows you to automatically log you into SOLIDWORKS websites, such as: MySolidWorks, SOLIDWORKS Forum,  Customer Portal, and Get Support.

*Figure-64. Login to Solidworks button*

# WORKFLOW IN SOLIDWORKS

The first step in SolidWorks is to create a sketch. After creating sketch of the desired feature, we create solid or surface model from that sketch. After doing the desired operations on the solid/surface model, we go for assembly or analyses. After, we are satisfied with the assembly/ analyses, we create the engineering drawings from the model to allow manufacturing of the model into a real world object.

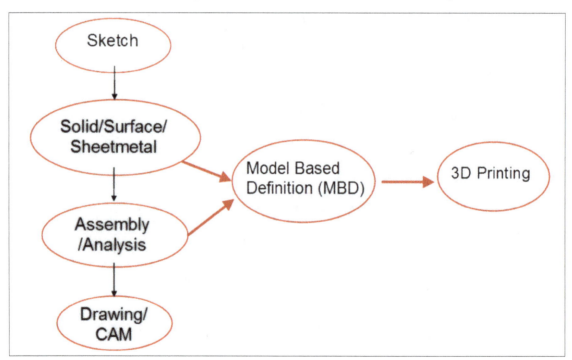

# SELF-ASSESSMENT

Q1. If you have downloaded the SolidWorks Setup files from Internet then the files will be available in Downloads folder of Windows by default. (T/F)

Q2. We cannot create the desktop icon of SolidWorks if we have not opted for it while installing. (T/F)

Q3. Status Bar in SolidWorks window also displays the tips related to current tool. (T/F)

Q4. Which of the following operation results in display of **New SolidWorks Document** dialog box?
(a)     Press CTRL+N
(b)     Click on **New** button from Menu Bar
(c)     Click on **New Document** link button from Task Pane.
(d)     All of the above

Q5. Which of the following is not an option in the **New SolidWorks Document** dialog box?
(a)     Part
(b)     Assembly
(c)     Drawing
(d)     Sketch

Q6.     The ............... **CommandManager** provides all modeling tools that are used for feature-based solid modeling.

Q7.     The ............. **CommandManager** is used to add dimensions and tolerances to the features of a part.

Q8.     The .......... **CommandManager** is used to insert a component and apply various types of mates to the assembly.

Q9.     Write down the steps to close the current document in SolidWorks.

Q10.   How can we change the unit system of current SolidWorks document?

Q11.   What is the purpose of Add-Ins in SolidWorks?

Q12. The .................... check box forces SolidWorks to use only fully defined sketches for creating features.

**Answer to Self-Assessment:**
**1.** T, **2.** F, **3.** T, **4.** d, **5.** d, **6.** Features, **7.** MBD, **8.** Assembly, **12.** Use fully defined sketches

# FOR STUDENT NOTES

# Chapter 2

# Sketching Practical and Practice

Topics Covered

The major topics covered in this chapter are:

- *Basic Sketching*
- *Advanced Sketching*

# PRACTICAL 1 SKETCHING (BASIC LEVEL)

Create sketch of a 2.5 mm thick plate as shown in Figure-1.

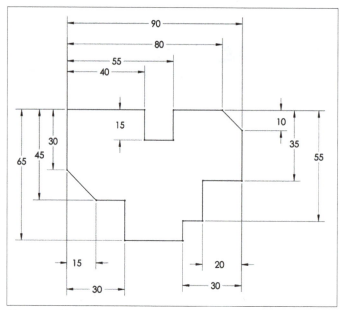

*Figure-1. Sketch for Practical 1*

Steps:

1. Start a new part file in SolidWorks.
2. Use the line tools to create sketch.

## Starting Sketch

- Start SolidWorks using Start menu or Desktop icon.
- Click on the **Part** button from **New** section in the **Welcome** dialog box; refer to Figure-2.

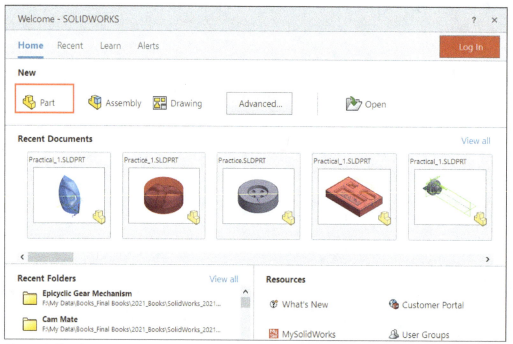

*Figure-2. Part button*

- Make sure **MMGS** option is selected in **Unit system** drop-down at the bottom right corner in the Application window; refer to Figure-3.

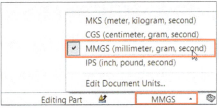

*Figure-3. MMGS unit option*

- Click on the **Sketch** tool from the **Sketch CommandManager** in the **Ribbon**. You will be asked to select a plane for creating sketch.
- Select the Front Plane from graphics area to create sketch. The sketching tools will be displayed.

## Settings for Sketching

- Click on the **Options** tool from the **Quick Access Toolbar**. The **System Options - General** dialog box will be displayed.
- Select the Sketch option from left in the dialog box and select following check boxes to create sketches efficiently.
  - Select the **Auto-rotate view normal to sketch plane on sketch creation and sketch edit** check box so that sketching plane automatically gets parallel to screen.
  - Select the **Use fully defined sketches** check box to allow exiting sketching environment only when all the geometries of sketch are dimensioned fully and leave nothing to guess. A fully defined sketch is necessary if you want to create manufacturable model.
  - Select the **Enable on screen numeric input on entity creation** check box to enter the size parameters for sketch entities when creating them.
  - Select the **Create dimension only when value is entered check box to automatically create dimensions when entered** in input boxes.
  - Click on the **Document Properties** tab in the dialog box and select **Dimensions** option from the left section. The parameters for defining dimension style will be displayed. Select **None** option from **Unit Precision** drop-down because our drawing has full number dimensions; refer to Figure-4.
  - Click on the **OK** button from the dialog box to apply settings.

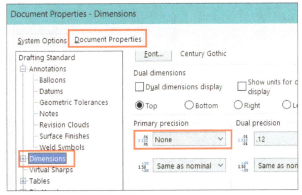

*Figure-4. Setting precision for dimension*

## Creating Line Sketch

- Click on the **Line** tool from **Line** drop-down in **Ribbon** or press **L** key (without **SHIFT** key). The **Insert Line PropertyManager** will be displayed at the left in Application window.
- Click at the Origin to specify start point of line and move the cursor towards right in horizontal direction.

Tip: Generally when you start sketching, the question arises where do we start the sketch. Answer to this question is based on dimensions created in the drawing. In Figure-5, most of the dimensions are referenced to top left corner so it would be easier to create sketch starting from there and move to the side which has maximum dimensions given which is towards right.

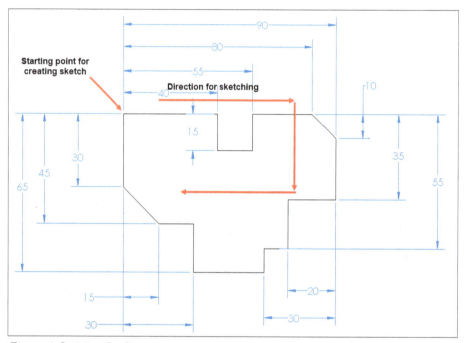

*Figure-5. Starting sketching*

- Specify **40** in the input box when preview of line is horizontal and press **ENTER**. The line will be created.
- Move the cursor downward in vertical line, specify **15** in input box and press **ENTER**. Now, move the cursor towards right in horizontal direction and click when distance is approximately 15; refer to Figure-6.

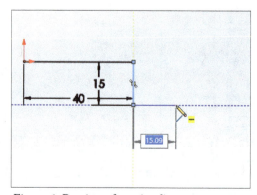

*Figure-6. Preview of creating line*

- Move the cursor upward and click when end point is horizontally collinear with previous 40 length line; refer to Figure-7.

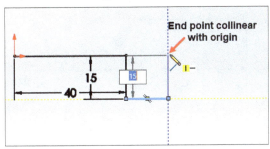

*Figure-7. Defining end point of vertical line*

- Move the cursor towards right in horizontal direction and click when length is approximately 25. Follow the procedure to create other entities of sketch; refer to Figure-8.

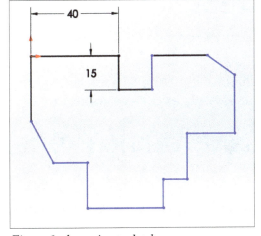

*Figure-8. Approximate sketch*

- Click on the **Smart Dimension** tool from **Ribbon**. You will be asked to select entities for dimensioning.
- Select the end points as shown in Figure-9 and place the dimension at proper distance from sketch. The **Modify** input box will be displayed.
- Specify the distance value as **55** and press **ENTER** to create the dimension.

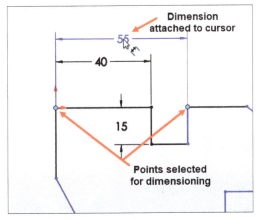

*Figure-9. Placing dimension*

- Similarly, place the other dimensions; refer to Figure-10.

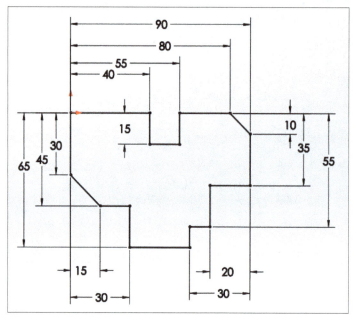

*Figure-10. Sketch after dimensioning*

- Click on the **Exit Sketch** tool from the **Ribbon** or top-right corner of graphics area.
- Press **CTRL+S** to save the sketch and save model in desired directory.

## PRACTICAL 2 SKETCHING (MODERATE LEVEL)

Create the sketch as shown in Figure-11.

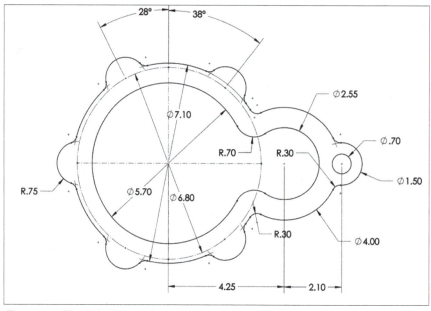

*Figure-11. Sketch for Practical 2*

Steps:
1. Start a new part file in SolidWorks.
2. Start sketch with IPS unit system.
3. Use circle, arc, and fillet tools.

### Creating Sketch

- Start SolidWorks using Start menu or Desktop icon.
- Create a new part model as discussed earlier.

- Select **IPS** option from the **Unit system** drop-down at the bottom-right corner of the application window.
- Start sketch on Front plane as discussed earlier.

Tip: When you have circles in sketch then most of the time we start creating sketch using center point of major circle at the origin and then everything is created with respect to that point.

- Click on the **Circle** tool from the **Sketch CommandManager** in the **Ribbon**. The **Circle PropertyManager** will be displayed and you will be asked to specify center point of the circle.
- Click at the origin to specify center point of circle. You will be asked to specify diameter of circle.
- Specify the diameter as **5.7** in input box and press **ENTER** to create circle.
- Similarly, create circles of diameter **7.10** and **6.80** at the origin; refer to Figure-12 and click on the **OK** button from the **PropertyManager**.

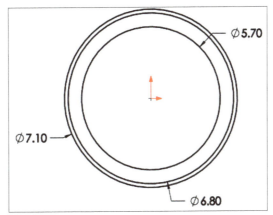

*Figure-12. Circles created*

- Select the circle of diameter **6.80** and select the **For construction** check box to make selected circle as construction circle; refer to Figure-13. Press **ESC** to exit selection and **PropertyManager**.

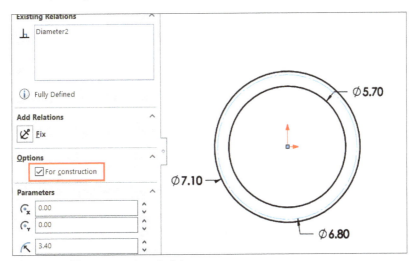

*Figure-13. For construction check box*

- Click on the **Centerline** tool from the **Line** drop-down in the **Sketch CommandManager** of **Ribbon**. The **Insert Line PropertyManager** will be displayed and you will be asked to specify start point for line.

- Create horizontal and vertical centerlines passing through center and quadrant points of construction circle; refer to Figure-14.

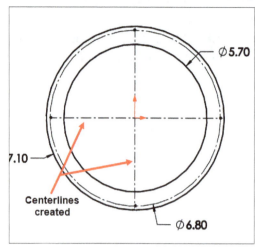

*Figure-14. Centerlines created*

- Click on the **Circle** tool from the **Sketch CommandManager** in the **Ribbon**. The **Circle PropertyManager** will be displayed and you will be asked to specify center point of circle.
- Create two circles of diameter 1.5 as shown in Figure-15.
- Click on the **Smart Dimension** tool from the **Ribbon** to apply angle dimensions for these circles. You will be asked to select entities for dimensioning.
- Select the points in the order as shown in Figure-16, the angle dimension will get attached to cursor.

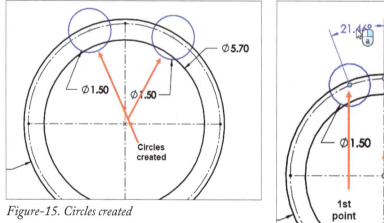

*Figure-15. Circles created*

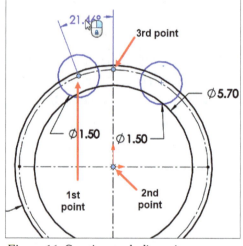

*Figure-16. Creating angle dimension*

- Click at desired location to place the dimension. You will be asked to specify the value of angle dimension.
- Type the value as **28** in the input box and press **ENTER**. The dimension will be applied. Similarly, apply 38$^0$ angle dimension to other circle; refer to Figure-17.

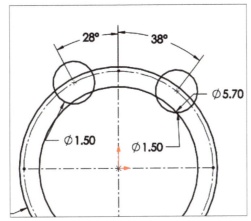

*Figure-17. Applying angle dimension*

- Press **ESC** to exit the tool.
- Select the two circles created at angles while holding the **CTRL** key and click on the **Mirror Entities** tool from **Sketch CommandManager** in the **Ribbon**. The **Mirror PropertyManager** will be displayed.
- Click in the **Mirror about** selection box and select horizontal construction line. Preview of mirror copy of circles will be displayed; refer to Figure-18. Click on the **OK** button from the **PropertyManager** to create the copy.

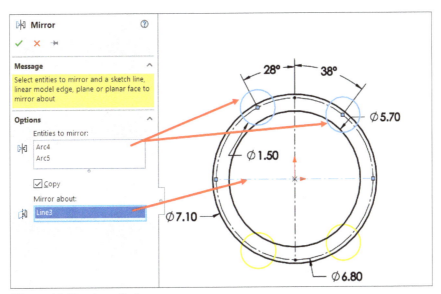

*Figure-18. Preview of mirror copy*

- Create the 5th smaller circle of diameter **1.5** at intersection of centerline and construction circle; refer to Figure-19.

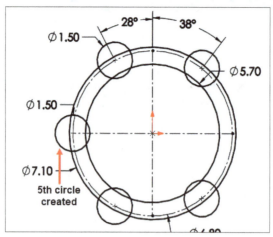

*Figure-19. Creating 5th small circle*

- Create other circles as shown in Figure-20 using **Circle** tool and **Smart Dimension** tool.

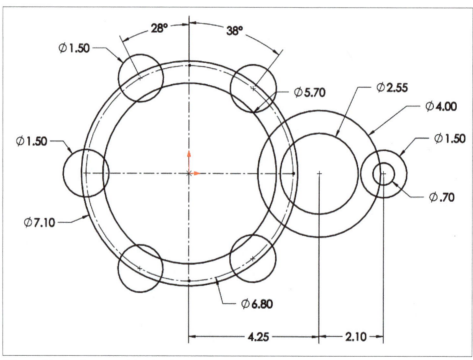

*Figure-20. Circles created for sketch*

- Click on the **Trim Entities** tool from the **Sketch CommandManager** in the **Ribbon** and trim extra portions of sketch by clicking and dragging cursor on them. If any of the dimensions/relations gets removed then reapply them. In our case, we had to apply equal constraints to all the smaller circles; refer to Figure-21 and apply dimension to circle of diameter 4.

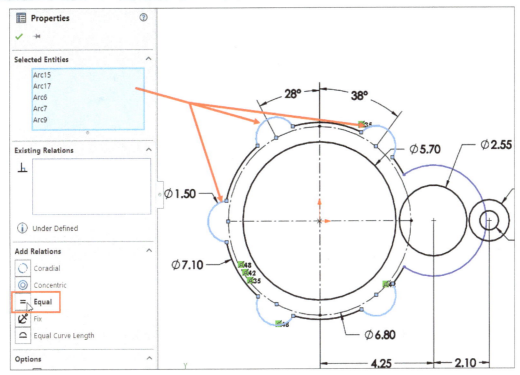

*Figure-21. Applying equal constraint*

- Click on the **Sketch Fillet** tool from the **Sketch CommandManager** in the **Ribbon**. The **Sketch Fillet PropertyManager** will be displayed.
- Specify value of radius as **0.3** in the **Fillet Radius** edit box of **PropertyManager** and hover the cursor at intersection point where fillet is to be created. Preview of fillet will be displayed; refer to Figure-22.
- Create other fillets in the sketch as per the drawing; refer to Figure-23. You can also select two intersecting arcs in place of intersection point to create fillet.

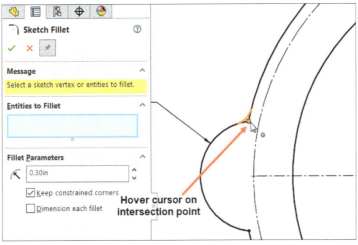

*Figure-22. Preview of fillet*

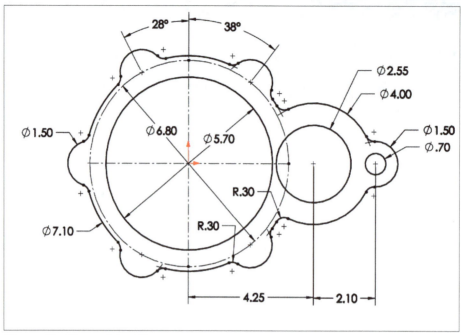

*Figure-23. After applying fillet*

- Now, only fillets with radius value 0.7 are left in sketch. Click on the **3 Point Arc** tool from the **Arc** drop-down in the **Ribbon**. The **Arc PropertyManager** will be displayed.
- Specify the end points of arc as shown in Figure-24. You will be asked specify radius of arc.

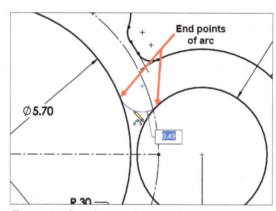

*Figure-24. Specifying end points of arc*

- Type **0.7** in the input box without moving mouse and press **ENTER**. The arc will be created. Press **ESC** to exit the **Arc PropertyManager**.
- Select the newly created arc and circle to apply tangent constraint. The **PropertyManager** as shown in Figure-25 will be displayed.
- Select the **Tangent** button from the **Add Relations** rollout in **PropertyManager** to apply constraint. Similarly, apply the tangent constraint between arc and other connected circle.
- Create mirror copy of this arc with respect to horizontal centerline as discussed earlier; refer to Figure-26.

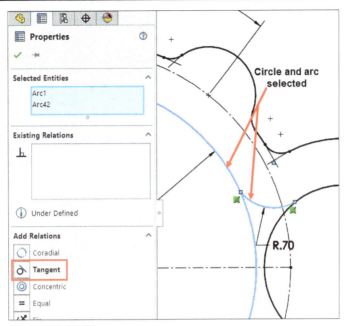

*Figure-25. Making arc tangent*

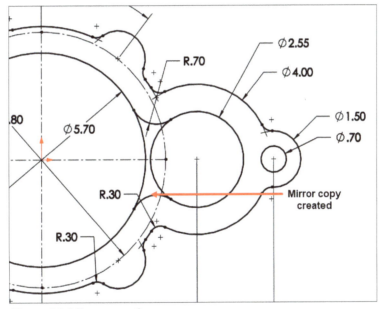

*Figure-26. Mirror copy of arc*

- Trim extra portions of circles to get the final sketch.

## PRACTICAL 3 SKETCHING (MODERATE LEVEL)

Create the sketch as shown in Figure-27.

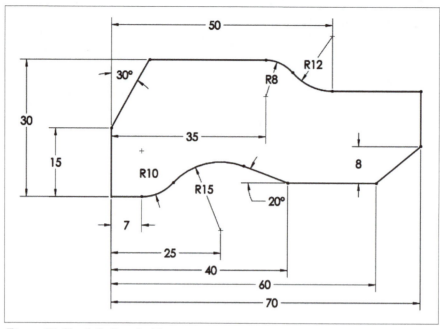

*Figure-27. Sketch for Practical 3*

Steps:
1. Start a new part file in SolidWorks.
2. Start sketch with MMGS unit system.
3. Use line, arc, and fillet tools.

## Creating Sketch

- Start SolidWorks using Start menu or Desktop icon.
- Create a new part model as discussed earlier.
- Set the unit to **MMGS** from the **Unit system** drop-down at the bottom-right corner of the application window.
- Click on the **Sketch** tool from the **Sketch CommandManager** in the **Ribbon**. You will be asked to select sketching plane.
- Select the **Front Plane** from graphics area to create sketch on front plane.
- Click on the **Line** tool from **Ribbon** and create vertical line of length **15** upward.
- Move the line inclined approximately 30 degree and click to create the line; refer to Figure-28.

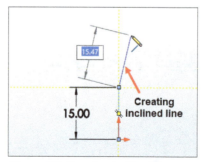

*Figure-28. Creating inclined line*

- Press **ESC** to exit the line tool and click on the **Smart Dimension** tool from the **Ribbon**. You will be asked to select entities for dimensioning.
- Select two lines as shown in Figure-29 to create angle dimension and click to place the dimension.

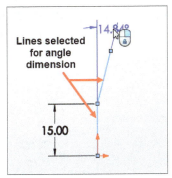

*Figure-29. Selecting lines for
angle dimension*

- Specify the angle value as **30** and press **ENTER** to apply dimension. Select two points as shown in Figure-30 and apply distance dimension as **30**.

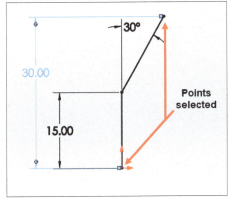

*Figure-30. Selecting points for dimensioning*

- Click on the **Line** tool and create a horizontal line of approximately 30 mm length.
- Now, move cursor back to origin, then towards right while simultaneously going downward. It will change line creation mode to tangent arc creation.

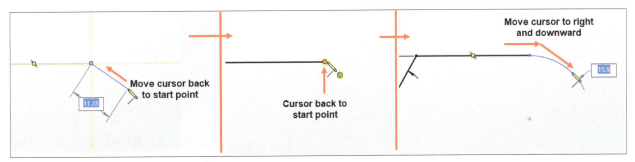

*Figure-31. Activating arc using line tool*

- Type the value as **8** in the input box and press **ENTER**. You will be asked to specify end point of arc.
- Click at an appropriate location as per the drawing to specify end point of arc; refer to Figure-32. The line creation mode will become active again.

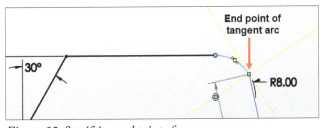

*Figure-32. Specifying end point of arc*

- Activate tangent arc creation mode again as discussed earlier; refer to Figure-33. You need to move cursor back to origin and then in the motion shown in Figure-33 with arrows.

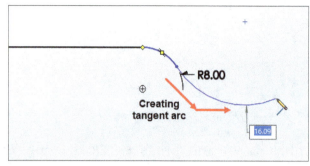

*Figure-33. Creating tangent arc*

- Enter the value of radius as **12** in the input box and click to specify end point as shown in Figure-34.
- Move the cursor towards right in horizontal line and click to specify end point when length of line is approximately **15**; refer to Figure-35.

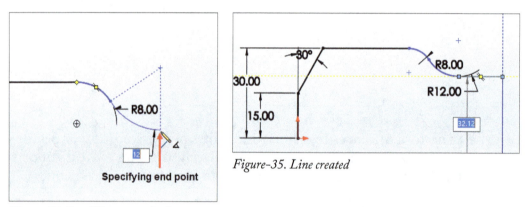

*Figure-35. Line created*

*Figure-34. Specifying arc end point*

- Create rest of the sketch entities as shown in Figure-36.

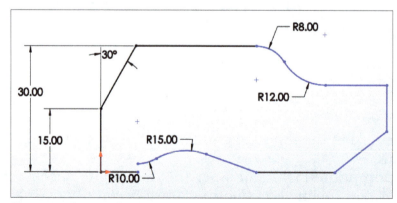

*Figure-36. Creating sketch entities*

- Select two end points while holding **CTRL** key and click on the **Merge** button to join the end points; refer to Figure-37.

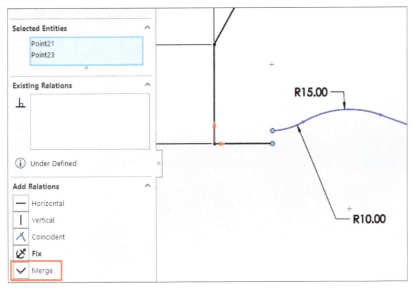

*Figure-37. Merging end points*

- Press **ESC** to exit selection.
- Apply the dimensions as shown in Figure-27.

## PRACTICAL 4 : SKETCHING (ADVANCED LEVEL)

Create the sketch as shown in Figure-38.

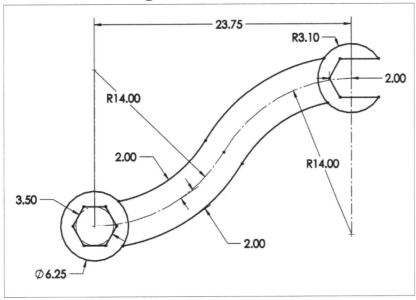

*Figure-38. Sketch for Practical 4*

Steps:
1. Start a new part file in SolidWorks.
2. Start sketch with MMGS unit system.
3. Use circle, arc, and polygon tools.

### Creating Sketch

- Start SolidWorks using Start menu or Desktop icon.
- Create a new part model as discussed earlier.
- Set the unit to **MMGS** from the **Unit system** drop-down at the bottom-right corner of the application window.

- Click on the **Sketch** tool from the **Sketch CommandManager** in the **Ribbon**. You will be asked to select sketching plane.
- Select the **Front Plane** from graphics area to create sketch on front plane.
- Click on the **Circle** tool from the **Sketch CommandManager** in the **Ribbon** and create a circle of diameter **6.25** at origin.
- Click on the **Polygon** tool from the **Ribbon** to create polygon. You will be asked to specify center point of polygon.
- Click at the center of circle to specify center of polygon and then specify end point of polygon in horizontal line with center point; refer to Figure-39.

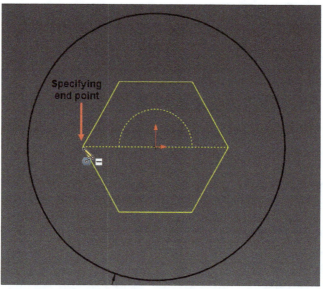

*Figure-39. Specifying end point of polygon*

- Click on the **Smart Dimension** tool from the **Ribbon** and create dimension as shown in Figure-40. Press **ESC** to exit the tool.

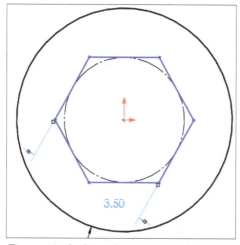

*Figure-40. Applying dimension to polygon*

- Select origin and the left side end point of polygon in horizontal line with origin; refer to Figure-41. Click on the **Horizontal** button to apply constraint.

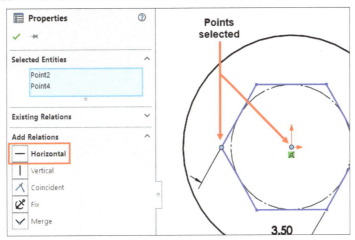

*Figure-41. Points for horizontal constraint*

- Click on the **Centerpoint Arc** tool from the **Arc** drop-down in the **Ribbon**. You will be asked to specify center point of arc.
- Create the arc as shown in Figure-42. Click on the **Tangent Arc** button from the **Arc PropertyManager** and create tangent arc as shown in Figure-43.

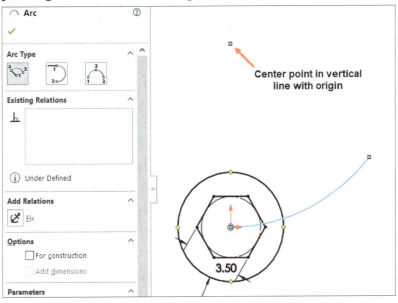

*Figure-42. Creating 1st arc*

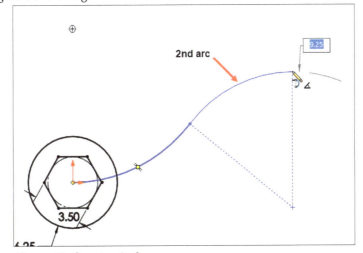

*Figure-43. Creating 2nd arc*

- Press **ESC** to exit the tool. Now, select two arcs created recently and select the **For construction** check box from the **Properties PropertyManager**; refer to Figure-44.

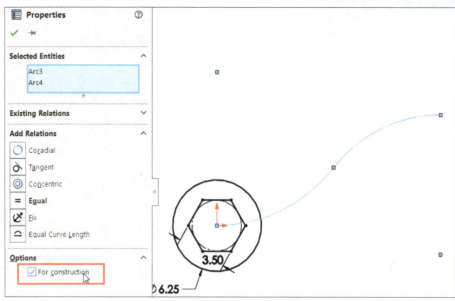

*Figure-44. For construction check box*

- Click on the **Smart Dimension** tool from the **Sketch CommandManager** in **Ribbon** and apply the dimensions as shown in Figure-45.

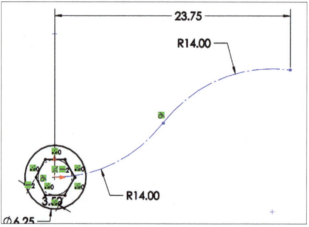

*Figure-45. Applying dimensions to arcs*

- Apply vertical constraint to end point of arc and center point as discussed earlier; refer to Figure-46.

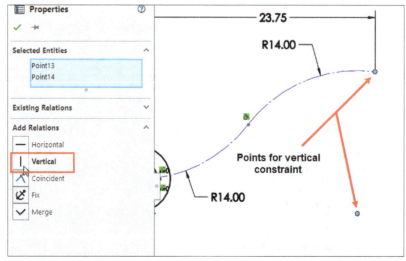

*Figure-46. Applying vertical constraint*

- Create circle of diameter 6.2 and hexagon at the end point of arc as shown in Figure-47. Also, apply dimensions and constraints as per the figure.

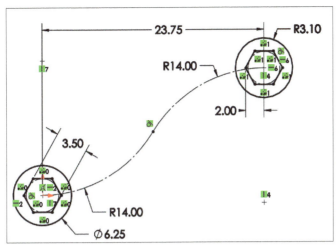

*Figure-47. Creating polygon and circle with dimensions applied*

- Click on the **Extend Entities** tool from the **Trim Entities** drop-down in the **Ribbon**. You will be asked to select entities to be extended.
- Select the lines of polygon to extend them; refer to Figure-48. Press ESC to exit the tool.

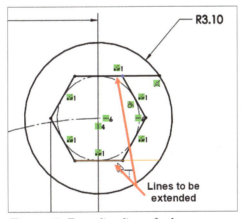

*Figure-48. Extending lines of polygon*

- Click on the **Trim Entities** tool from the **Ribbon** and remove extra segments of sketch; refer to Figure-49.

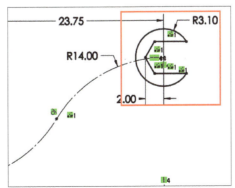

*Figure-49. After trimming segments*

- Select two construction arcs from graphics area and click on the **Offset Entities** tool from the **Ribbon**. The **Offset Entities PropertyManager** will be displayed.

- Set the parameters as shown in Figure-50 and click on the **OK** button from **PropertyManager**.

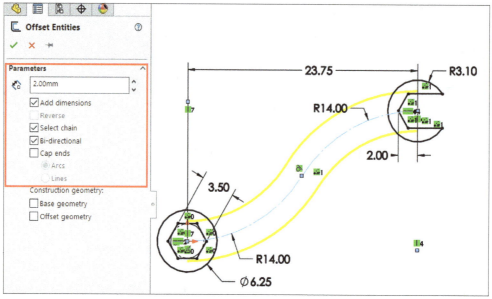

Figure-50. *Applying offset to construction arcs*

- Trim extra portions of sketch to get final sketch.

## PRACTICE 1 TO 9

Create the sketches given in Figure-51 to Figure-59.

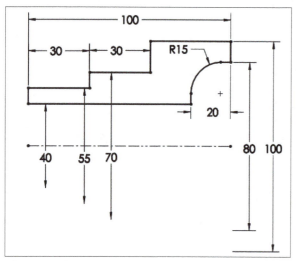

Figure-51. *Sketch for Practice 1*

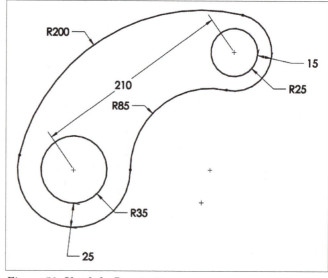

Figure-52. *Sketch for Practice 2*

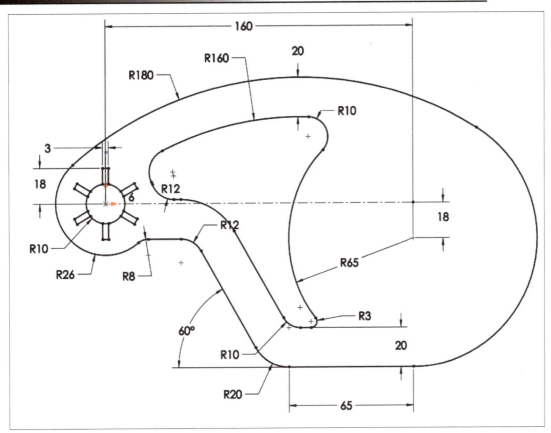

*Figure-53. Sketch for Practice 3*

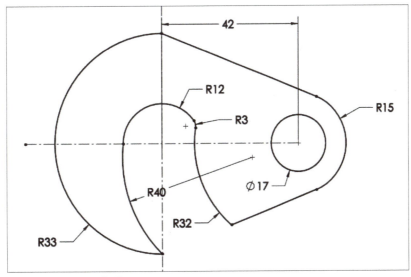

*Figure-54. Sketch for Practice 4*

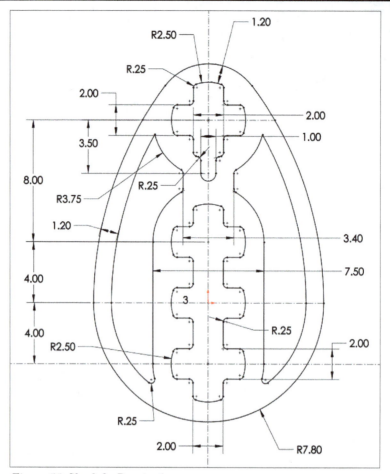

*Figure-55. Sketch for Practice 5*

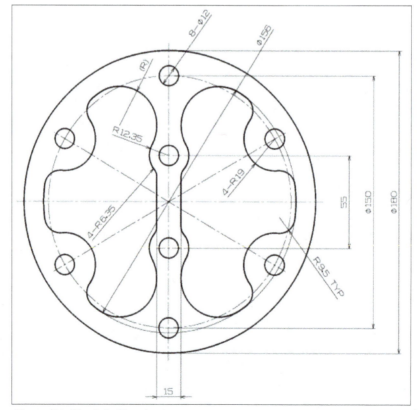

*Figure-56. Sketch for Practice 6*

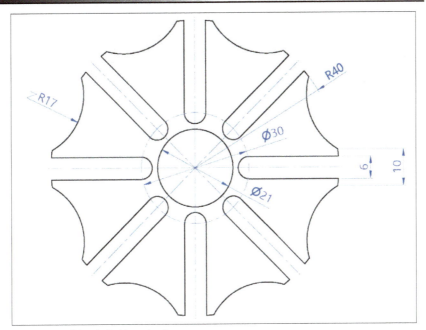

*Figure-57. Sketch for Practice 7*

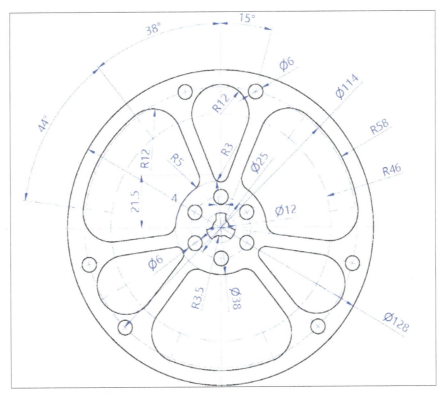

*Figure-58. Sketch for Practice 8*

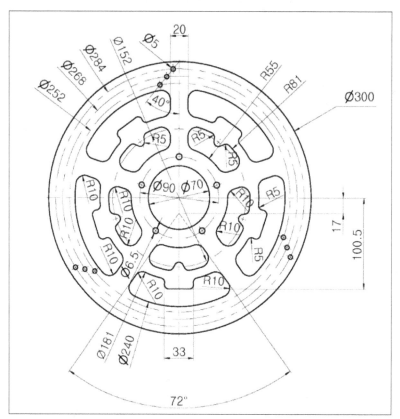

*Figure-59. Sketch for Practice 9*

FOR STUDENT NOTES

# FOR STUDENT NOTES

# Chapter 3

# Solid Modeling
# Practical and Practice

## Topics Covered

The major topics covered in this chapter are:

- *Basic Solid Modeling*
- *Advanced Solid Modeling*

## PRACTICAL 1 (BASIC LEVEL)

Create 3D Model using drawing shown in Figure-1.

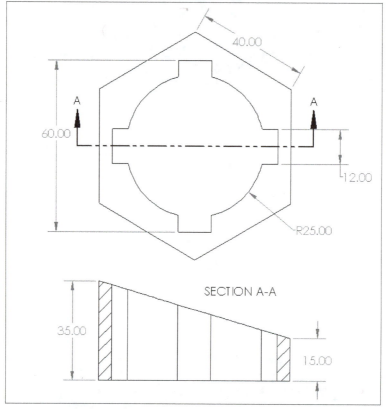

*Figure-1. Solid Modeling Practical 1*

## Observations on Drawing / Blueprint Reading

The outer boundary of model is hexagon of size 40 mm. Now, circle inside the hexagon can be a boss feature or cut feature by looking only at the top view of drawing. But when you check the section view, it becomes clear that circle is a cut feature otherwise it would have been protruding above the outer hexagon boundaries. The same can be said about the rectangular key cuts attached to the circle. Try to visualize the model in your mind and compare whether you reach the result shown in at the end of this practical.

Steps:
- Create sketch given in the top of view of model on Top view. Why on top view? Because it is easier to create model when you match top view, right view, and front view of model with Top plane, Right plane, and Front plane; respectively. Also, model is oriented based on selected planes.
- Create extrude feature of the sketch up to 35 mm because in section view, the walls are straight.
- After creating extrude feature, you need to create an extrude cut feature that cuts the model in taper with one side at 35 mm and other side at 15 mm.

## Creating Sketch

- Start a new part model document in SolidWorks.
- Click on the **Sketch** tool from the **Sketch CommandManager** in the **Ribbon**. You will be asked to select a sketching plane.
- Select the Top Plane from graphics area. Sketching environment will become active.

- Create the sketch using **Polygon**, **Circle**, **Center Rectangle**, and **Trim Entities** tools; refer to Figure-2.

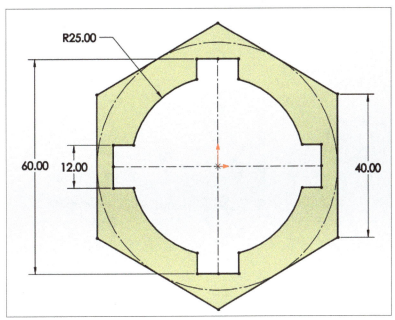

*Figure–2. Practical 1 base sketch*

## Creating Extrude and Extrude Cut Features

- Click on the **Extruded Boss/Base** tool from the **Features CommandManager** in the **Ribbon**. The **Extrude PropertyManager** will be displayed.
- Select any geometry of the sketch from graphics area. The **Boss-Extrude PropertyManager** will be displayed.
- Expand **Selected Contours** rollout and select the sketch region as shown in Figure-3. Make sure to delete any other sketch region from the selection box.

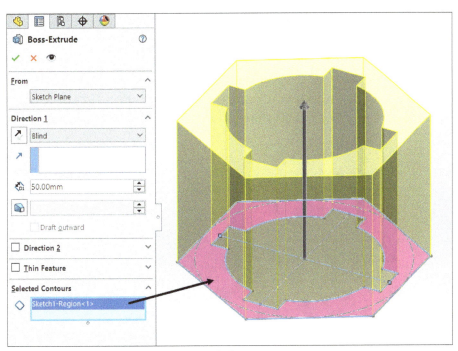

*Figure–3. Selecting sketch section for extrude feature*

- Specify the height of extrude feature as **35** in the **Depth** edit box of **Direction 1** rollout in the **PropertyManager** and click on the **OK** button.

- Click on the **Extruded Cut** tool from the **Features CommandManager** in the **Ribbon**. You will be asked to select a plane for creating sketch of extruded cut feature.
- Select the Front Plane from **Model Tree** in graphics area; refer to Figure-4 to create sketch because section of model is displayed in front view of the drawing.

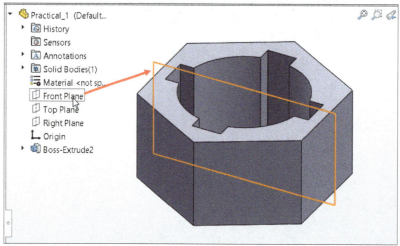

*Figure–4. Selecting plane for cut feature*

- Create line sketch (triangle) as shown in Figure-5 for extruded cut feature. Click on the **Exit Sketch** tool from the **Ribbon** after creating sketch. Preview of feature will be displayed.

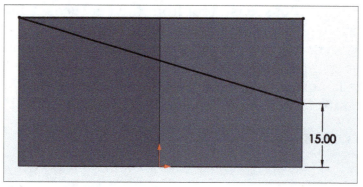

*Figure–5. Sketch for cut feature*

- Select the **Through All -Both** option from the **End Condition** drop-down in the **Direction 1** rollout of **PropertyManager**. Preview of cut feature will be displayed; refer to Figure-6.

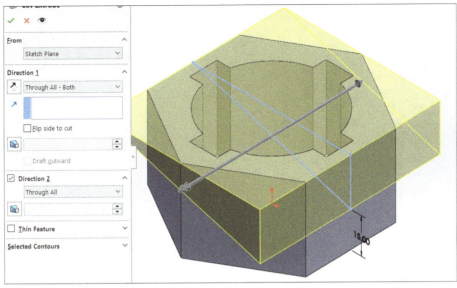

*Figure-6. Preview of cut feature*

- Click on the **OK** button from the **PropertyManager** to create cut feature. The model will be displayed as shown in Figure-7.

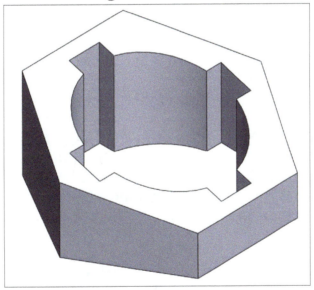

*Figure-7. Model for Practical 1*

## PRACTICAL 2 (BASIC LEVEL)

Creating the solid model using drawing shown in Figure-8.

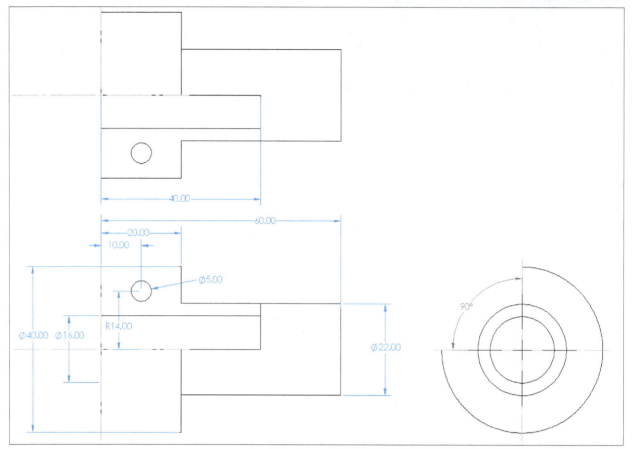

*Figure-8. Drawing for Practical 2*

## Observations on Drawing / Blueprint Reading

By looking at the drawing, it can be easily visualized that the model is a shaft with outer diameter of 22 mm and inner diameter of 16 mm. The shaft also has a flange of diameter 40 mm with a hole of diameter 5 mm. After creating shaft with flange, a revolve cut is created removing material in 90 degree span.

Steps:
- Create sketch given at bottom in drawing (which is front view) on Front plane using outer boundary of sketch in view.
- Create revolve feature of 360 degree span using the sketch.
- Create sketch for revolve cut feature on Top plane and create revolve cut feature up to a span of 90 degree.

## Creating Revolve Feature

- Start SolidWorks and then a new part file with MMGS unit system.
- Click on the **Sketch** tool from the **Sketch CommandManager** in the **Ribbon**. You will be asked to select sketching plane.
- Select the Front Plane from graphics area. The sketching environment will become active.
- First, create centerline passing through origin and then create rest of the sketch using lines; refer to Figure-9.

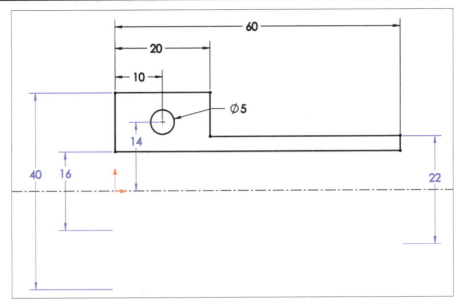

*Figure-9. Sketch for revolve feature*

- Click on the **Revolved Boss/Base** tool from the **Features CommandManager** in the **Ribbon**. The **Revolve PropertyManager** will be displayed.
- Select the centerline from graphics area. Preview of revolve feature will be displayed; refer to Figure-10.

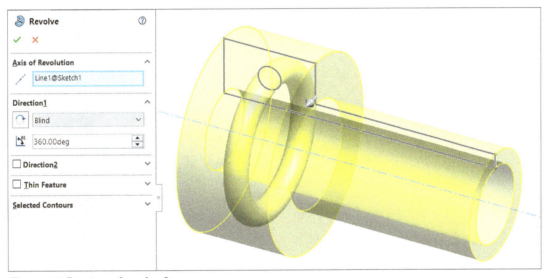

*Figure-10. Preview of revolve feature*

- Make sure **360 deg** is specified in the **Direction 1 Angle** edit box of **PropertyManager** and click on the **OK** button to create revolve feature.

## Creating Revolve Cut Feature

- Click on the **Revolved Cut** tool from the **Features CommandManager** in the **Ribbon**. The **Revolve PropertyManager** will be displayed.
- Select the **Top Plane** from **Model Tree** and create sketch for revolve cut feature as shown in Figure-11.

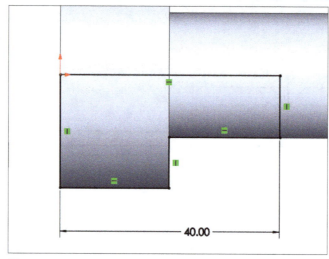

*Figure-11. Sketch for revolve cut feature*

- Click on the **Exit Sketch** tool from the **Ribbon**. The **Cut-Revolve PropertyManager** will be displayed.
- Select the centerline of sketch and set parameters for revolve cut as shown in Figure-12.

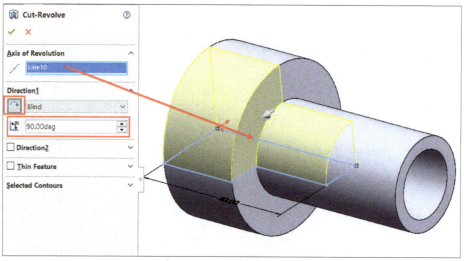

*Figure-12. Preview of revolve cut feature*

- Click on the **OK** button from the **PropertyManager**. The model will be created as shown in Figure-13.

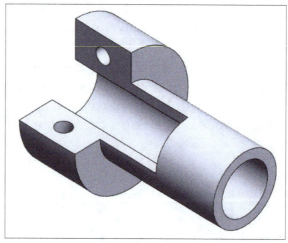

*Figure-13. Model for Practical 2*

## PRACTICAL 3 (MODERATE LEVEL)

Create 3D model based on drawing given in Figure-14. Assume the missing dimensions.

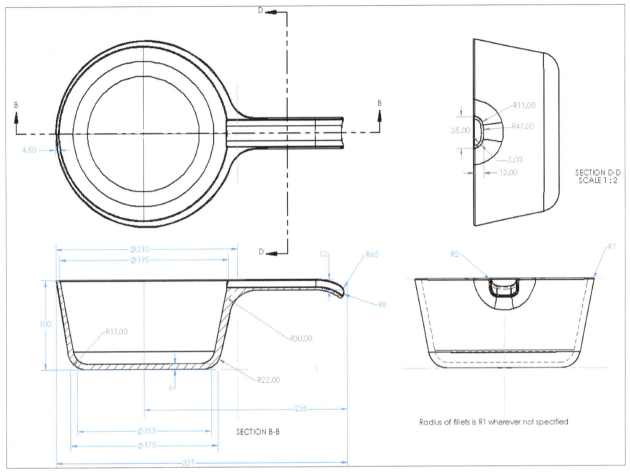

*Figure-14. Drawing for Practical 3*

## Observations on Drawing / Blueprint Reading

By looking at the drawing, first line of thoughts will be to create extrude feature with taper faces but taper angle is not provided in drawing. In drawing, we are provided two diameters for defining upper and lower faces of the pan for both inner and outer walls. Also, the thickness of walls is not uniform which means, we need to create loft features: one will be protrusion using outer diameter values and another will be a loft cut feature removing material from the protrusion feature. Handle of the pan is curved at end point so it would be easier to use sweep protrusion feature to create it.

Steps:
- Create circle of diameter 170 at top plane and circle of diameter 210 on a plane 100 mm distant from top plane upward. Using these circles, create loft feature.
- Similarly, create circle of diameter 155 at the top plane and circle of diameter 195 on the plane 100 mm distance from top plane upward. Create a loft cut feature to remove material from previous feature.
- Create sketch for sweep feature based on Section D:D in drawing and use path based on Section B:B drawing.
- Apply fillets based on the drawing.

## Creating Loft Feature

- Start SolidWorks if not started yet and create a new part file.
- Click on the **Sketch** tool from the **Sketch CommandManager** in the **Ribbon**. You will be asked to select sketching plane.
- Select the Top Plane from graphics area and create two circles as shown in Figure-15.

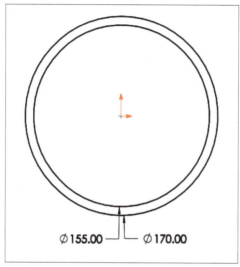

*Figure-15. Sketch on Top plane*

- Click on the **Plane** tool from **Reference Geometry** drop-down in **Features CommandManager** in the **Ribbon**.
- Select **Top Plane** from **Model Tree** in graphics area and specify distance as **100** in the edit box; refer to Figure-16. Click on the **OK** button from **PropertyManager** to create the plane.

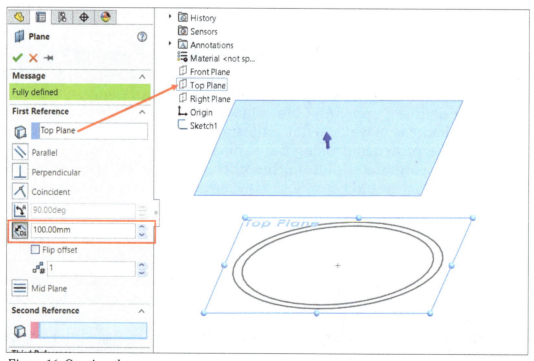

*Figure-16. Creating plane*

- Select newly created plane and two circles as shown in Figure-17.

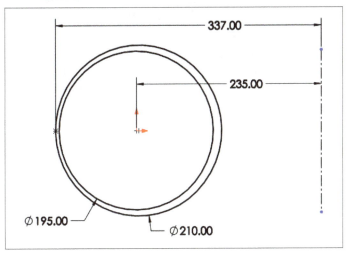

*Figure-17. Sketch created on plane*

- Click on the **Lofted Boss/Base** tool from the **Features CommandManager** in the **Ribbon**. The **Loft PropertyManager** will be displayed.
- Select the outer circles on top plane and offsetted plane. Preview of loft feature will be displayed; refer to Figure-18. Click on the **OK** button to create the feature.

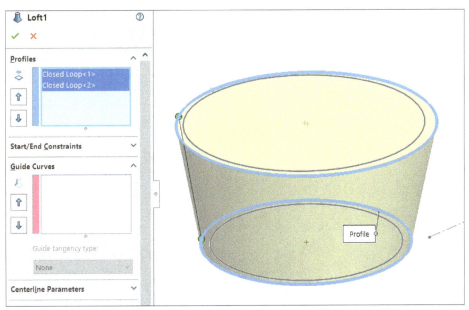

*Figure-18. Preview of loft feature*

## Creating Loft Cut Feature

- Select two sketches under Loft1 feature in the **FeatureManager Design Tree** and select **Show** button from mini toolbar; refer to Figure-19.

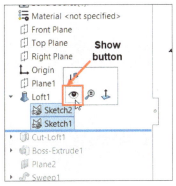

*Figure-19. Show button*

- Click on the **Lofted Cut** tool from the **Features CommandManager** in the **Ribbon**. The **Cut-Loft PropertyManager** will be displayed.
- Select the internal circles and keep the handles in straight line. Preview of loft cut feature will be displayed; refer to Figure-20. Click on the **OK** button from the **PropertyManager** to create the feature; refer to Figure-21.

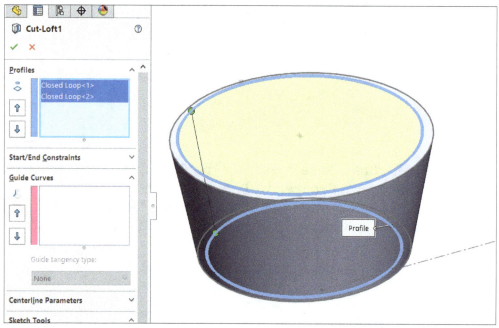

*Figure-20. Preview of loft cut feature*

*Figure-21. After creating loft cut feature*

## Creating Extrude Feature

- Click on the **Extruded Boss/Base** tool from the **Features CommandManager** in the **Ribbon** and select the outer circle of sketch created on top plane; refer to Figure-22.
- Specify the depth of extrude feature as **6** and click on the **OK** button to create the feature.

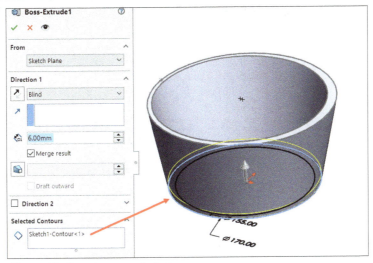

*Figure-22. Creating extrude feature*

## Creating Sweep Feature for Handle

- Start a new sketch on Front Plane and create line & arc for sweep path as shown in Figure-23.

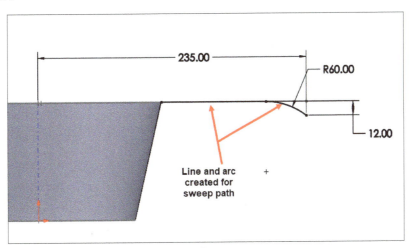

*Figure-23. Sketch created for sweep path*

- Create a new plane at the end point of line create in the previous sketch, parallel to Right Plane; refer to Figure-24.

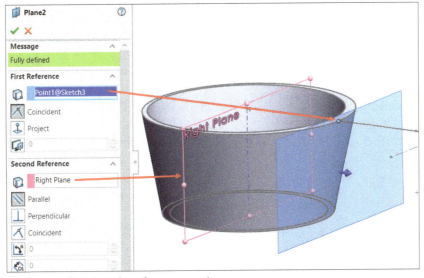

*Figure-24. Creating plane for sweep section*

- Create the sketch for section of sweep feature on newly created plane as shown in Figure-25.

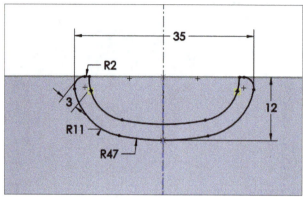

*Figure-25. Section sketch for sweep*

- Click on the **Swept Boss/Base** tool from the **Features CommandManager** in the **Ribbon** and set the parameters as shown in Figure-26. Make sure the **Align with end faces** check box is selected so that sweep feature is extended up to round face of pan.

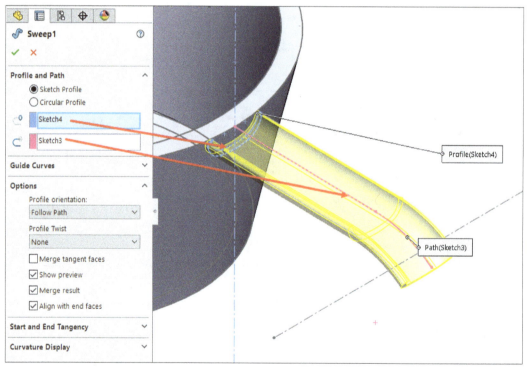

*Figure-26. Creating handle*

## Creating Extrude Cut Feature

- Create the plane as shown in Figure-27.
- Create the extrude cut feature using **Extruded Cut** tool as shown in Figure-28.

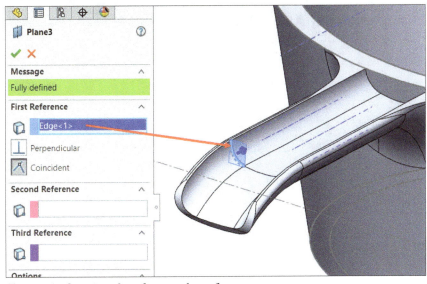

*Figure–27. Creating plane for extrude cut feature*

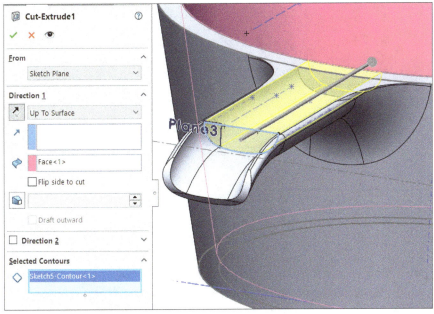

*Figure–28. Creating extruded cut feature*

- Apply fillets to the model to get final model; refer to Figure-29.

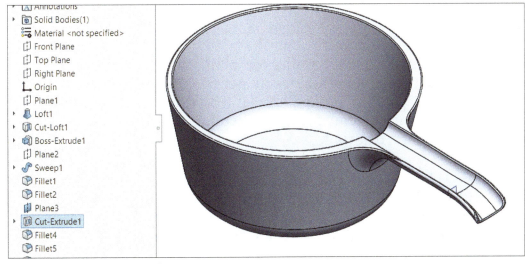

*Figure–29. Model for Practical 3*

# PRACTICAL 4 (MODERATE LEVEL)

Create 3D model based on drawing given in Figure-30.

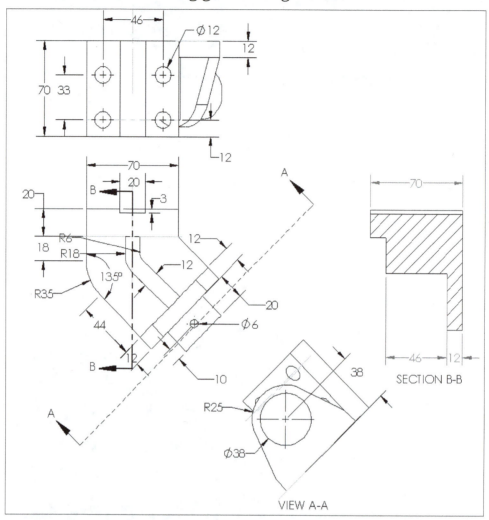

*Figure-30. Drawing for Practical 4*

## Observation on Drawing/Blueprint Reading

Most of the elements in this drawing are boss features that can be easily created by extrude features. The center wall of thickness 12 mm and height 46 mm can be created by using Rib feature. Using fillet tool, we can apply radius to faces as per drawing.

Steps:
* Create the base extrude feature using top plane up to thickness of 12 mm.
* Create extrude features on both sides as per the drawing.
* Create sketch for rib feature using plane at a distance of 46 from top face of base feature and create the rib feature using sketch.
* Apply fillets as per the drawing.

## Create Base Feature

* Start SolidWorks if not started yet and create a new part file document with MMKS unit system.
* Click on the **Sketch** tool from the **Sketch CommandManager** in the **Ribbon**. You will be asked to select a plane for creating sketch.
* Select the Top Plane from graphics area and create sketch as shown in Figure-31.

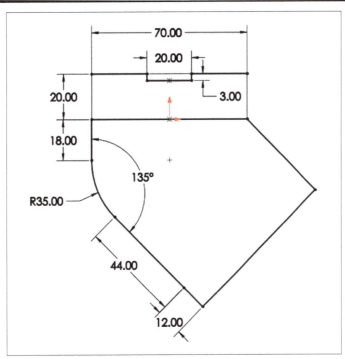

*Figure-31. Sketch for base feature*

- Click on the **Extruded Boss/Base** tool from the **Features CommandManager** in the **Ribbon**. The **Extrude PropertyManager** will be displayed.
- Select any boundary curve of the sketch. Preview of thin feature will be displayed.
- Clear the **Thin Feature** check box, clear the contours from **Selected Contours** selection box and select the region as shown in Figure-32. Preview of extrude feature will be displayed.

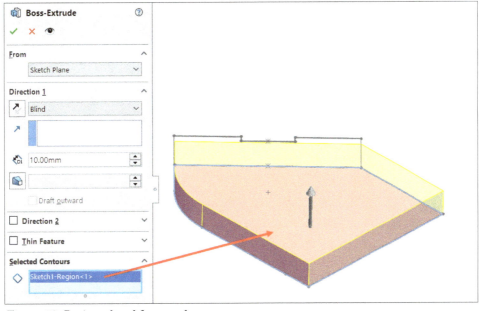

*Figure-32. Region selected for extrude*

- Specify the depth of extrude feature as **12** and click on the **OK** button to create the feature.

## Creating First Side Wall

- Select the Sketch1 feature from Boss-Extrude1 feature in **Model Tree** and **Show** button from mini toolbar; refer to Figure-33. The sketch will become visible.

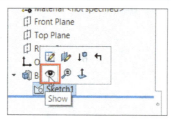

*Figure-33. Show button*

- Click on the **Extruded Boss/Base** tool from **Features CommandManager** in the **Ribbon** and select boundary as shown in Figure-34.

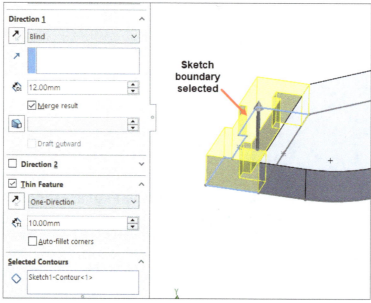

*Figure-34. Sketch boundary selected*

- Clear the **Thin Feature** check box, clear selection box of **Selected Contours** rollout and select the region as shown in Figure-35.

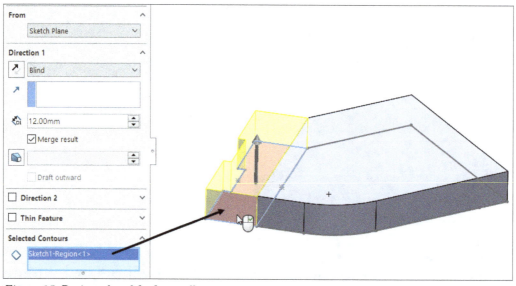

*Figure-35. Region selected for first wall*

- Specify the depth of wall as **70** in the **Depth** edit box and click on the **OK** button to create the feature.

## Creating Second Side Wall

- Click on the **Extruded Boss/Base** tool from the **Ribbon** and select the face as shown in Figure-36 to use a sketching plane.

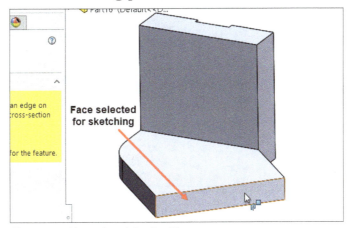

*Figure-36. Face selected for sketching*

- Create the sketch as shown in Figure-37 and extrude it to 12 mm backward.

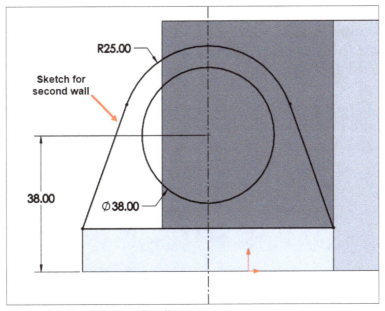

*Figure-37. Sketch for second wall*

- Make sure to select all the regions in sketch; refer to Figure-38. After setting parameters, click on the **OK** button from **PropertyManager** to create the feature.

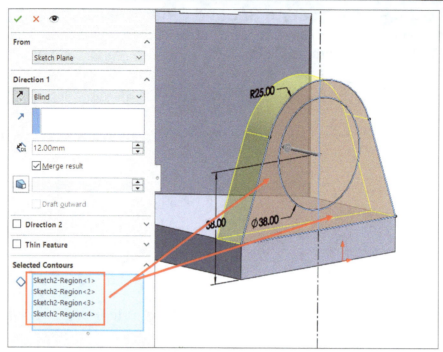

*Figure-38. Regions selected for second wall*

- Set the sketch of third extrude feature to **Show** as discussed earlier.
- Click on the **Extruded Boss/Base** tool from **Ribbon** and select the inner circle of sketch.
- Set the depth of feature as **20** in the **Depth** edit box of **PropertyManager**; refer to Figure-39 and create the feature.

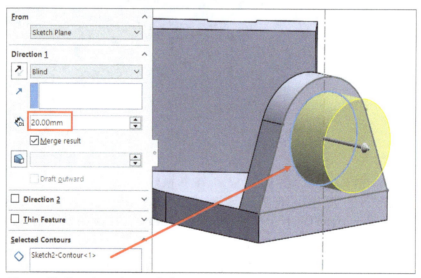

*Figure-39. Creating 4th extrude feature*

## Creating Rib Feature

- Click on the **Plane** tool from **Reference Geometry** drop-down in the **Features CommandManager** of the **Ribbon**. The **Plane PropertyManager** will be displayed.
- Select the top face of base extrude feature created earlier and specify distance as 46 mm; refer to Figure-40.

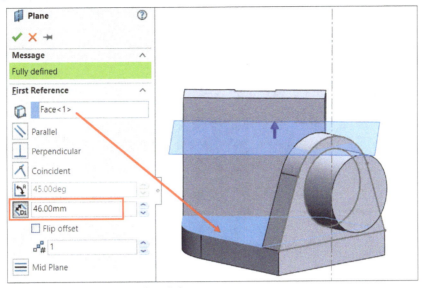

*Figure-40. Creating plane for rib feature*

- After specifying parameters, click on the **OK** button from **PropertyManager**. The plane will be created.
- Select the newly created plane and click on the **Sketch** tool from **Sketch CommandManager** in the **Ribbon**. The sketching plane will become parallel to screen.
- Create the sketch as shown in Figure-41.

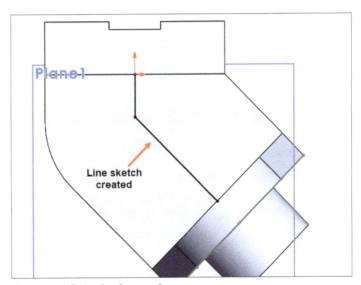

*Figure-41. Line sketch created*

- Click on the **Rib** tool from **Features CommandManager** in the **Ribbon** and select newly created sketch line. Preview of rib feature will be displayed; refer to Figure-42.

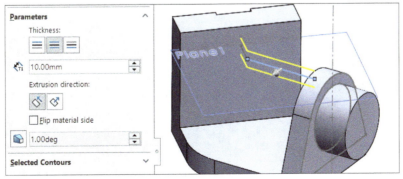

*Figure-42. Preview of rib feature*

• Click on the **Normal to Sketch** button from the **Extrusion direction** section of **PropertyManager**. Set the parameters as shown in Figure-43 and click on the **OK** button to create the rib feature; refer to Figure-44.

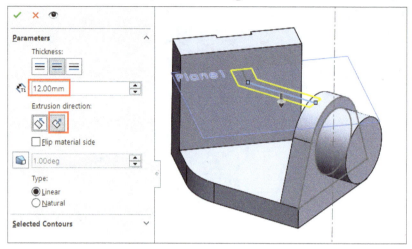

*Figure-43. Preview of rib feature*

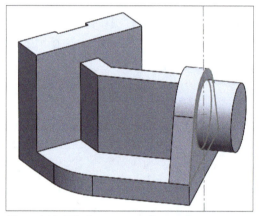

*Figure-44. Rib feature created*

## Creating Holes on First Wall

• Click on the **Extruded Cut** tool from the **Features CommandManager** in the **Ribbon**. You will be asked to select face/plane for sketching.
• Select the face of first wall; refer to Figure-45 and create the sketch as shown in Figure-46.

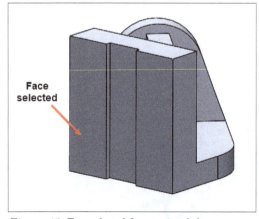

*Figure-45. Face selected for creating holes*

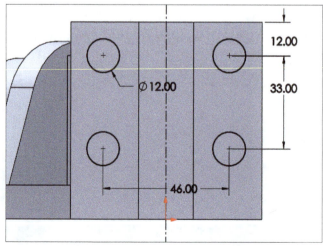

*Figure-46. Sketch created for holes*

- Click on the **Exit Sketch** tool from the **Sketch CommandManager** in the **Ribbon** and set the parameters as shown in Figure-47.

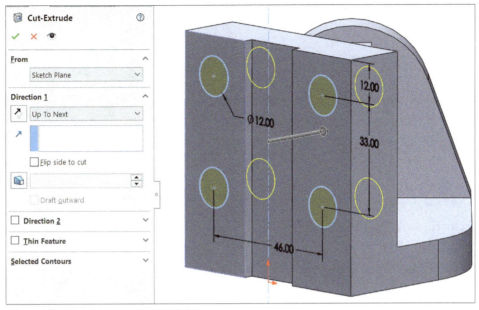

*Figure–47. Creating extruded cut holes*

- Click on the **OK** button from **Cut-Extrude PropertyManager** to create the holes.

## Creating Hole on Second Wall

- Click on the **Plane** tool from the **Reference Geometry** drop-down in the **Features CommandManager** of the **Ribbon**. The **Plane PropertyManager** will be displayed.
- Create the plane tangent to round face of second wall and parallel to top face of rib feature; refer to Figure-48.

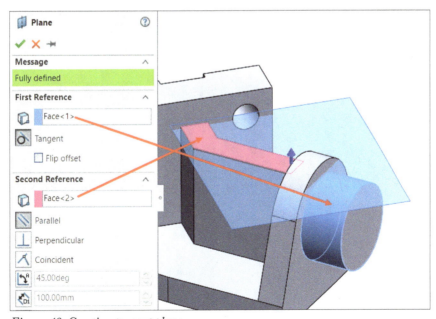

*Figure–48. Creating tangent plane*

- Create a vertical hole through the round boss feature as discussed earlier; refer to Figure-49 and Figure-50.

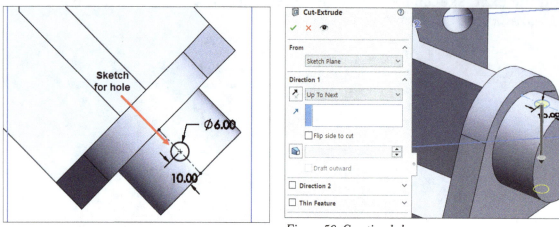

Figure-49. Sketch for hole

Figure-50. Creating hole

- Apply the fillets as per the drawing.

## PRACTICAL 5 (ADVANCED LEVEL)

Creating the model using drawing given in Figure-51 and Figure-52.

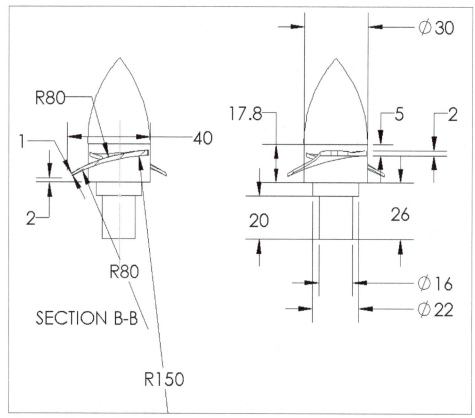

Figure-51. Practical 5 drawing

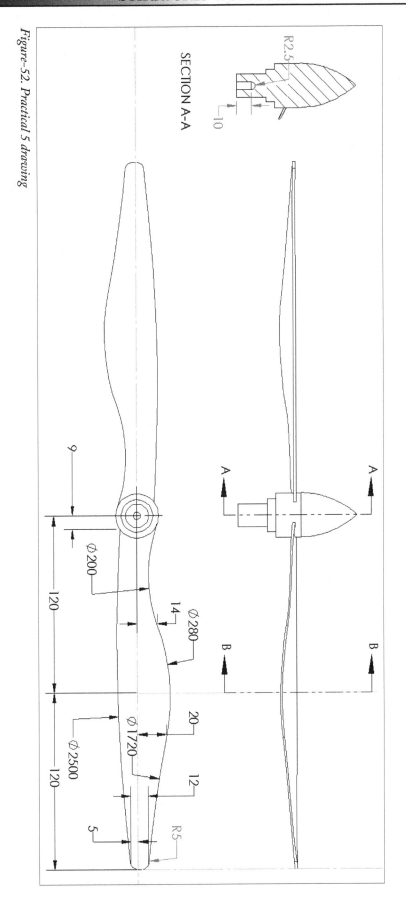

SECTION A-A

Figure-52. Practical 5 drawing

## Observation on Drawing/Blueprint Reading

The model of turbine blade is a single model but for modeling purpose, we can assume it as combination of two bodies: the center hub of turbine blade is created by using revolve feature and blade is created by using loft feature with extrude cut. Why blade is created using loft feature and not extrude feature? Because the blade is twisted at the mid while both ends of blade are in flat plane. This design can be achieved by using multiple sections at different locations joined by a protrusion feature Loft in SolidWorks. After creating loft feature, we will need a material removal feature to get final shape of blade. So, we will use extrude cut feature using boundary curve of blades.

Steps:
- Create the base revolve feature using Right plane based on drawing given in Figure-51.
- Create section sketches for loft feature on Right plane, plane at 120 from Right plane, and other plane at 240 from Right plane.
- Create loft feature using section sketches.
- Create sketch for extrude cut feature on Top plane and create the cut feature removing material from loft feature.

## Creating Base Revolve Feature

- Start SolidWorks if not started yet and create a new part document.
- Click on the **Revolved Boss/Base** tool from the **Features CommandManager** in the **Ribbon**. The **Revolve PropertyManager** will be displayed and you will be asked to select a sketching plane.
- Select the **Right Plane** from graphics area. Sketching mode will activate and tools of **Sketch CommandManager** will be displayed in **Ribbon**.
- Create the half section of center hub of turbine blade with centerline as shown in Figure-53. After creating sketch, click on the **Exit Sketch** tool from **Ribbon**. Preview of revolve feature will display. Specify the parameters as shown in Figure-54 and click on the **OK** button from **PropertyManager** to create the feature.

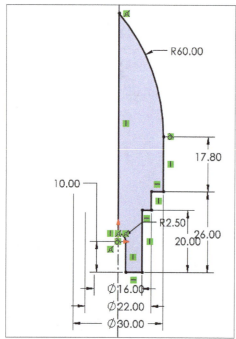

*Figure-53. Sketch for revolve feature*

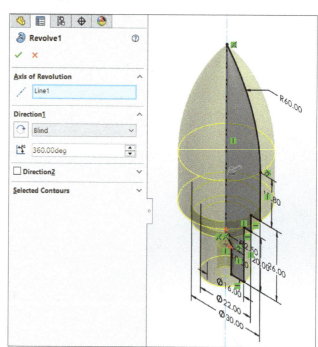

*Figure-54. Preview of revolve feature*

## Creating Loft Feature

- Before creating loft feature, we will need three section sketches at a distance of 120 mm from each other. So, we will first create planes to be used as base for these sketches. Click on the **Plane** tool from **Reference Geometry** drop-down in **Features CommandManager** of the **Ribbon**. The **Plane PropertyManager** will be displayed.
- Select the **Right Plane** from graphics area or **FeatureManager Design Tree**; refer to Figure-55. The preview of offset plane will be displayed.

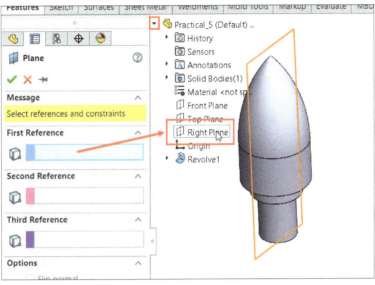

*Figure-55. Right plane*

- Specify the offset distance as **120** mm in the **Offset distance** edit box and **2** in the **Number of planes to create** edit box. Preview of planes will be displayed; refer to Figure-56.

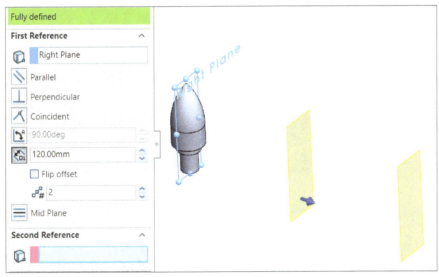

*Figure-56. Preview of planes*

- Click on the **OK** button from **PropertyManager** to create planes.
- Select the **Right Plane** from **FeatureManager Design Tree** and click on the **Sketch** tool from **Sketch CommandManager** in the **Ribbon**. The sketching environment will become active.

- Create the sketch as shown in Figure-57 on Right Plane. Similarly, create the sketches as shown in Figure-58 and Figure-59 on planes at a distance of 120 mm and 240 mm from Right Plane, respectively.

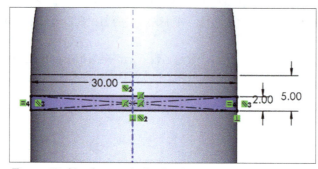

*Figure-57. Sketch created of right plane*

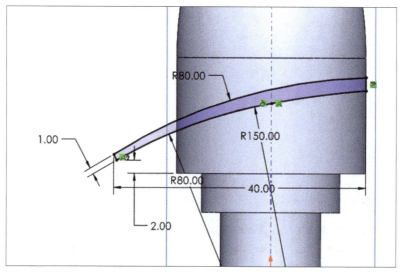

*Figure-58. Sketch at 120mm plane*

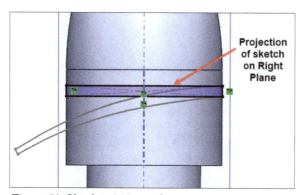

*Figure-59. Sketch at 240mm plane*

- Click on the **Lofted Boss/Base** tool from the **Features CommandManager** in the **Ribbon**. The **Loft PropertyManager** will be displayed.
- Select the sketches in sequence as shown in Figure-60. Note that you need to click on locations near green dots in Figure-60 so that the feature does not get twisted. Preview of loft feature will be displayed.

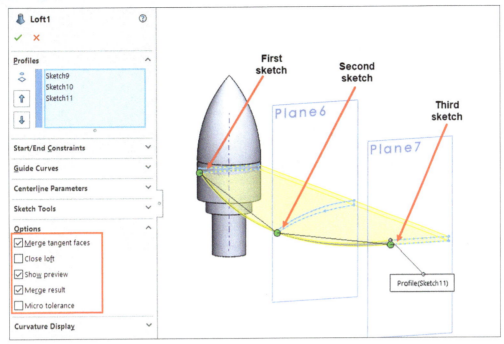

*Figure-60. Selecting sketch sections for loft*

- Click on the **OK** button from the **PropertyManager** to create the feature and hide the planes recently created.

## Creating Extrude Cut Feature

- Click on the **Extruded Cut** tool from the **Features CommandManager** in the **Ribbon**. The **Extrude PropertyManager** will be displayed and you will be asked to select plane for creating base sketch.
- Select the Top Plane from graphics area or **FeatureManager Design Tree** and create the sketch as shown in Figure-61.

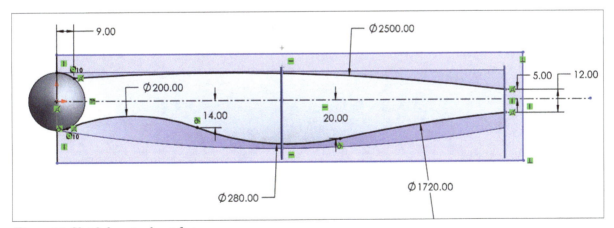

*Figure-61. Sketch for extrude cut feature*

- After creating sketch, click on the **Exit Sketch** tool from the **Ribbon** and specify the parameters as shown in Figure-62.

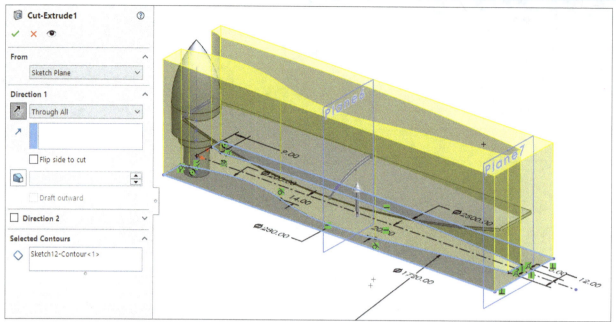

*Figure-62. Preview of extrude cut feature*

- Click on the **OK** button from **PropertyManager** to create the extrude cut feature.

## Creating Pattern for Mirroring Blade

- We are not using **Mirror** tool to create mirror copy of blade because that will make twisted section of blade on same side which is not required in our case. So, we will use the **Circular Pattern** tool for creating mirror copy. Click on the **Circular Pattern** tool from the **Linear Pattern** drop-down in **Features CommandManager** of the **Ribbon**. The **Circular Pattern PropertyManager** will be displayed.
- Select the blade from graphics area and specify the parameters as shown in Figure-63.

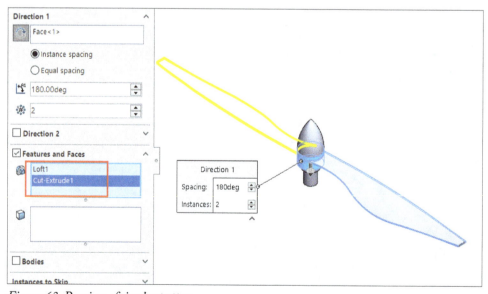

*Figure-63. Preview of circular pattern*

- Click on the **OK** button from the **PropertyManager** to create the feature.
- Apply the fillets of radius 5 mm at vertical edges of blades; refer to Figure-64.

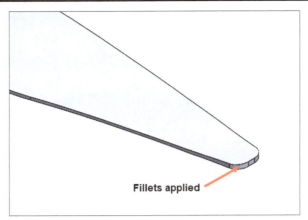

*Figure-64. Fillets applied*

# PRACTICAL 6 (ADVANCED LEVEL)

Create the model as shown in Figure-65.

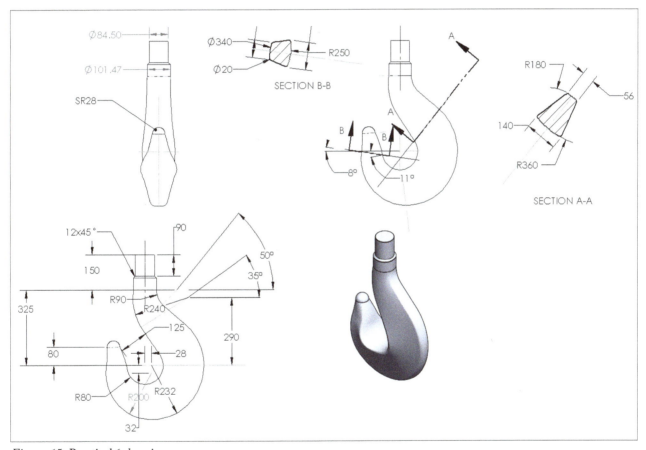

*Figure-65. Practical 6 drawing*

## Observation on Drawing/Blueprint Reading

The model of hook is a single model but for modeling purpose, we can assume it as combination of three bodies: the top part of hook is a cylinder created as revolved feature, mid section of hook is solid protrusion created by using different sections so it can be assumed as loft feature, and at the end point of hook is a semi sphere which can be created using revolved feature.

Steps:
• Create the guide curves of loft feature on Front Plane.
• Create reference planes at different locations as per the drawing.

- Create sketch sections at those planes and then use **Lofted Boss/Base** tool to create the feature.
- Using **Revolved Boss/Base** tool, create the top and end sections of hook.

## Creating Loft Feature

- Create sketch for guide curves on Front plane as shown in Figure-66.

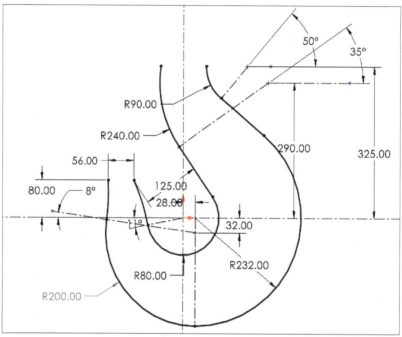

*Figure-66. Sketch for guide curves*

- Click on the **Plane** tool from the **Reference Geometry** drop-down in the **Features CommandManager** of the **Ribbon**. The **Plane PropertyManager** will be displayed and you will be asked to select reference objects.
- Select the arc with radius 90 and the point at the start on adjacent curve; refer to Figure-67. Preview of plane will be displayed. Click on the **OK** button to create plane.

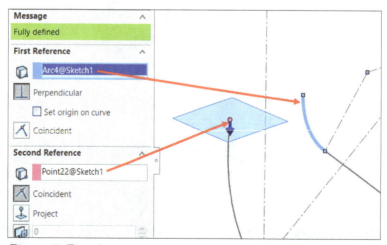

*Figure-67. First plane*

- Similarly, create other planes; refer to Figure-68, Figure-69, Figure-70, and Figure-71.

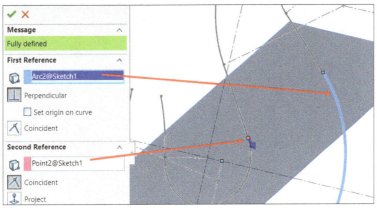

Figure-68. Second plane

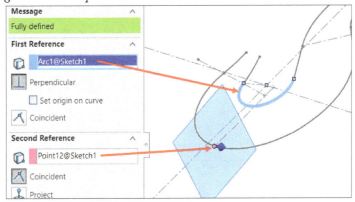

Figure-69. Third plane

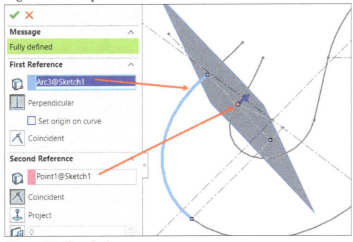

Figure-70. Fourth plane

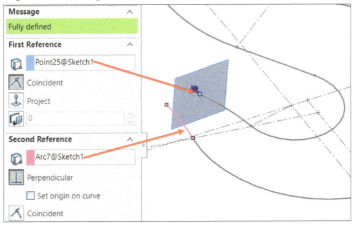

Figure-71. Fifth plane

- Create sketches on these planes as per the figures given next.

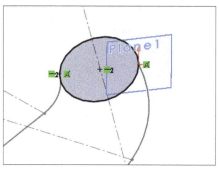

*Figure-72. Sketch on plane 1*

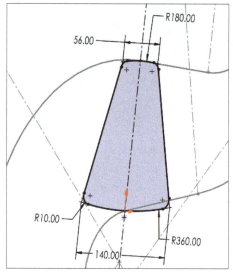

*Figure-73. Sketch on plane 2*

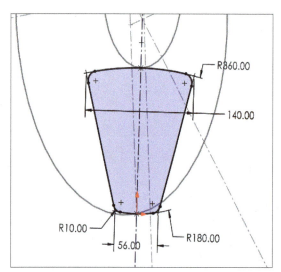

*Figure-74. Sketch on plane 3*

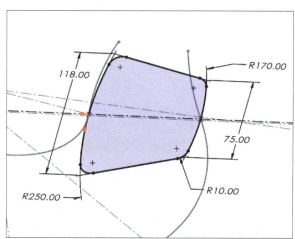

*Figure-75. Sketch on plane 4*

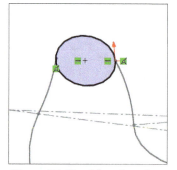

*Figure-76. Sketch on plane 5*

- After creating these sketches on respective planes, click on the **Loft** tool from **Features CommandManager** in the **Ribbon**. The **Loft PropertyManager** will be displayed.
- Select the sketch sections in sequence and then select the guide curves; refer to Figure-77. Note that connectors should be on similar locations at different section sketches when you are selecting the sections.

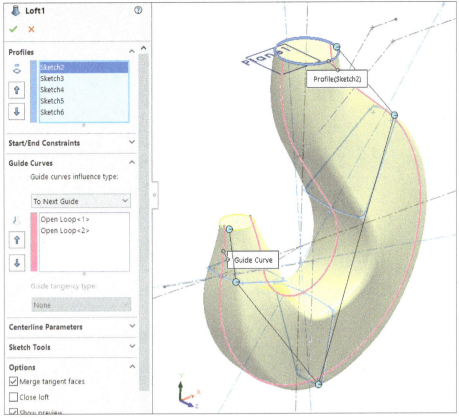

*Figure-77. Preview of loft feature*

- Click on the **OK** button from **PropertyManager** to create the feature.

## Creating Revolve Features

- Click on the **Revolved Boss/Base** tool from the **Features CommandManager** in the **Ribbon** and select the flat face at the end of hook as sketching plane; refer to Figure-78. Sketching environment will become active.
- Create semi-circle using boundary of flat surface; refer to Figure-79.

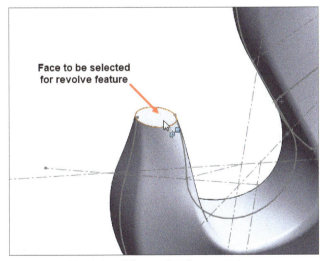

*Figure-78. Face selected for revolve feature section*

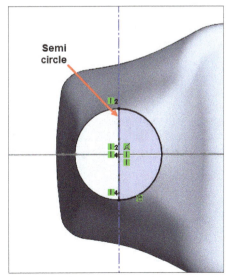

*Figure-79. Semi circle created for revolve feature*

- After creating sketch, click on the **Exit Sketch** button from **Ribbon**. Preview of revolve feature will be displayed. Set the angular span as 180 degree and other parameters as shown in Figure-80.

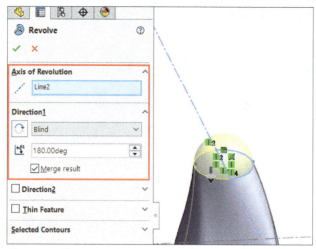

*Figure-80. Preview of revolved feature*

- Click on the **OK** button from **PropertyManager** to create the feature.
- Similarly, create revolved feature on other end of hook; refer to Figure-81.

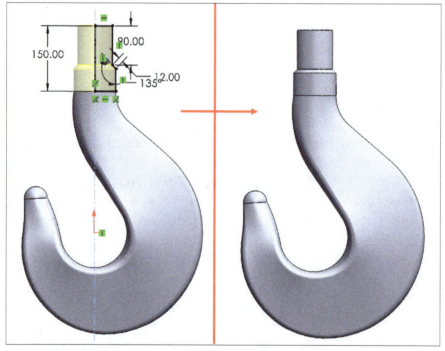

*Figure-81. Second revolved feature*

## PRACTICE 1 TO 10

Create solid models using drawings shown in figures given next.

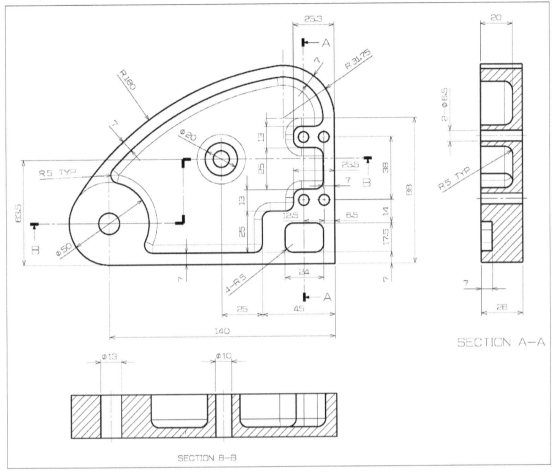

*Figure-82. Solid Modeling Practice 1*

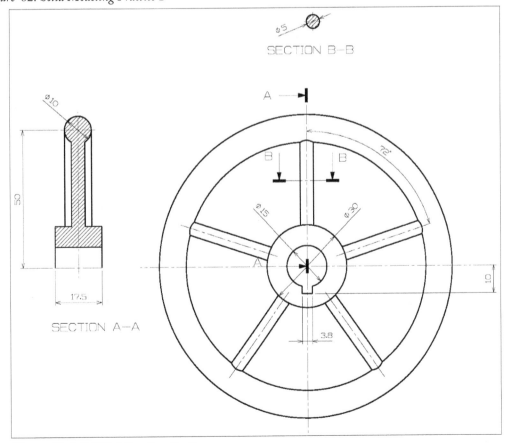

*Figure-83. Solid Modeling Practice 2*

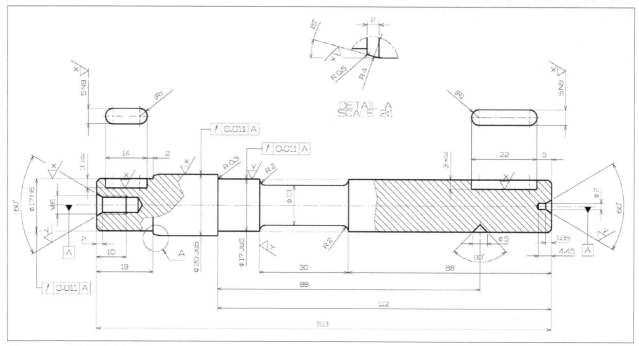

*Figure-84. Solid Modeling Practice 3*

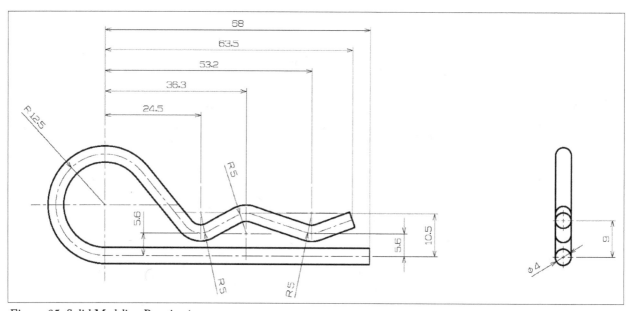

*Figure-85. Solid Modeling Practice 4*

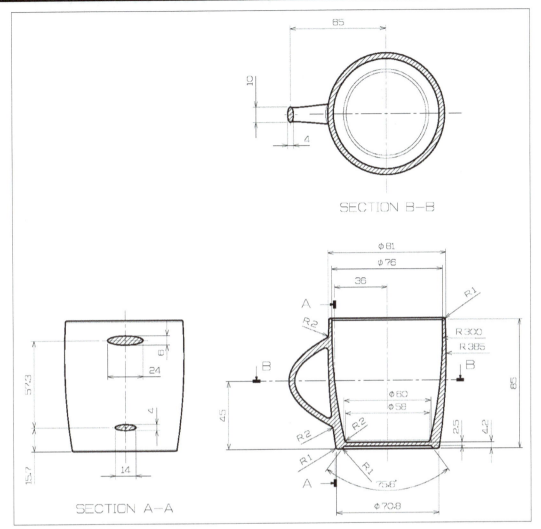

Figure-86. Solid Modeling Practice 5

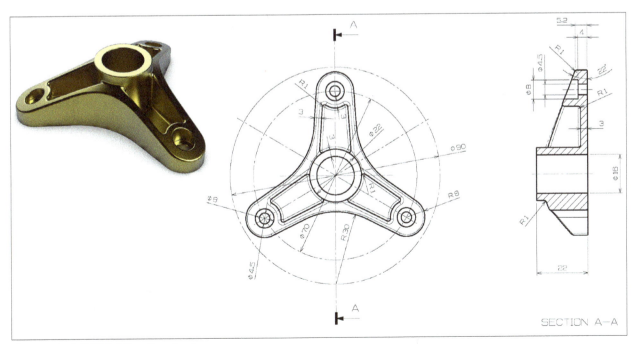

Figure-87. Solid Modeling Practice 6

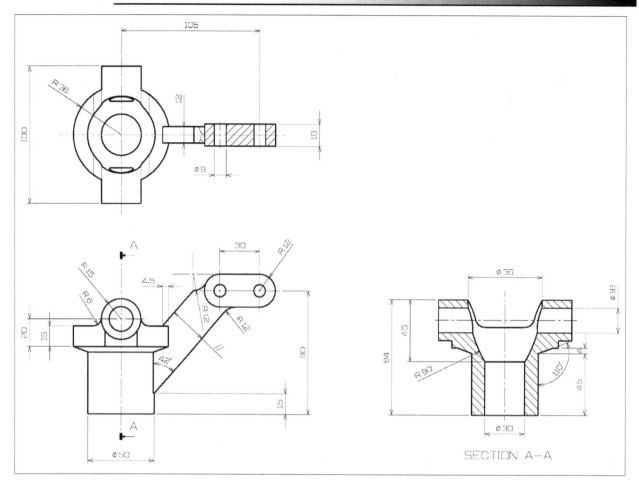

*Figure-88. Solid Modeling Practice 7*

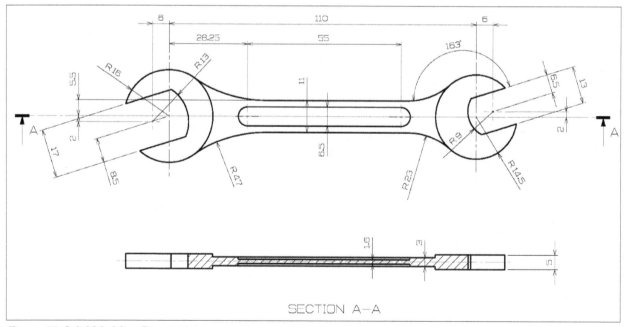

*Figure-89. Solid Modeling Practice 8*

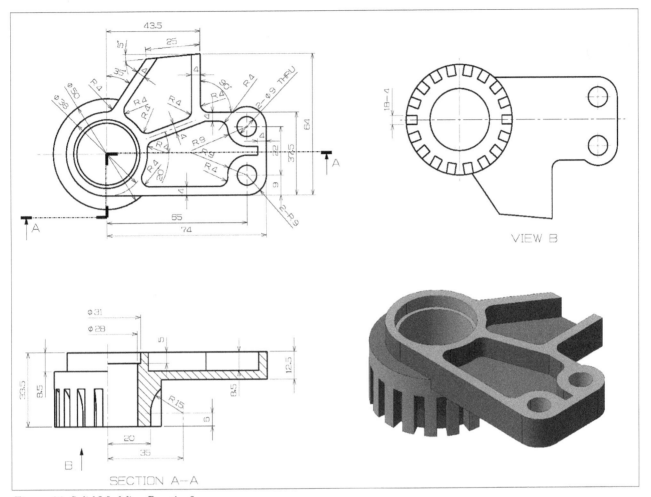

*Figure-90. Solid Modeling Practice 9*

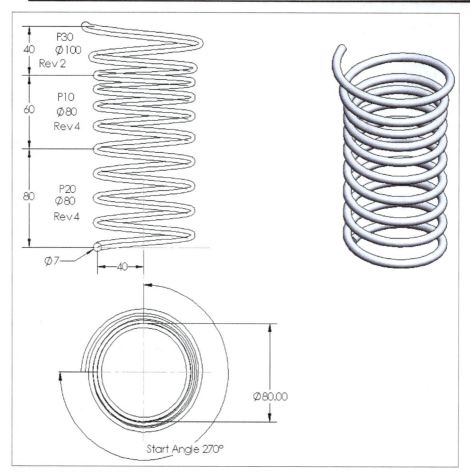

*Figure-91. Solid Modeling Practice 10*

# FOR STUDENT NOTES

FOR STUDENT NOTES

# Chapter 4

# Surface Modeling
# Practical and Practice

## Topics Covered

The major topics covered in this chapter are:

- *Basic Surface Modeling*
- *Advanced Surface Modeling*

# SCOPE OF SURFACE MODELING

Surface modeling is reserved for complex shapes to be manufactured like aesthetically designed water bottles, showpieces, outer shell of automobiles and aeroplanes to get better aerodynamics, and so on. After creating surface model, you will add thickness to the model so that it can be manufactured. You will find major use of surface modeling in plastic and sheet metal industries.

# PRACTICAL 1 (BASIC SURFACE MODEL)

Create the surface model as shown in Figure-1 and apply thickness of 2 mm to the model.

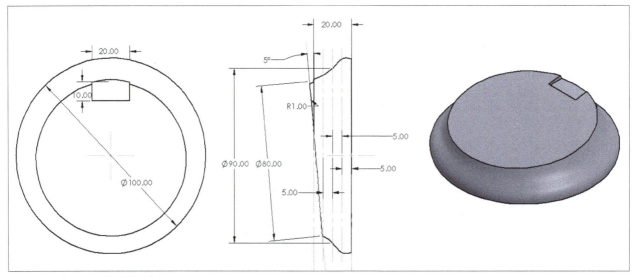

*Figure-1. Surface Modeling Practical 1*

## Observation of Drawing / Blueprint Reading

The model do not have a thickness so clearly it can be created using surface modeling. The walls of model can be created using loft feature and flat faces can be created using filled surface. Before creating loft feature, you need to create planes as per the drawing and then create sketches on them. To create indent on surface, you need to create extrude surface and then use trim tools. After performing trimming, knit the surfaces to form one body to which thickness can be applied.

## Starting Document and Creating Sketches

- Start SolidWorks if not started yet and create a new Part document.
- Click on the **Plane** tool from the **Reference Geometry** drop-down in the **Features CommandManager** of **Ribbon**. The **Plane PropertyManager** will be displayed.
- Select the Top Plane from graphics area and specify offset distance as 5 mm; refer to Figure-2.
- Similarly, create Plane 2 and Plane 3 at 5 mm consecutive distances; refer to Figure-3.

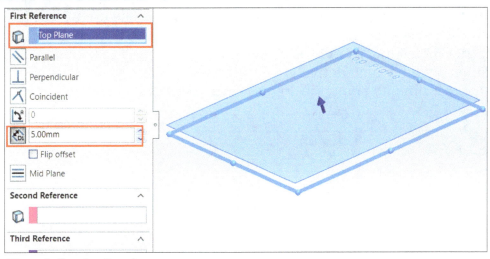

*Figure-2. Creating first plane*

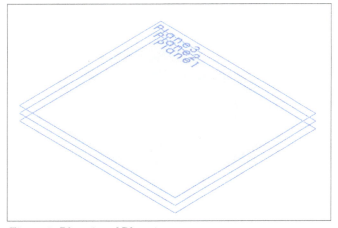

*Figure-3. Plane 2 and Plane 3*

• Create the sketches on Top Plane, Plane 1, and Plane 2 as shown in Figure-4.

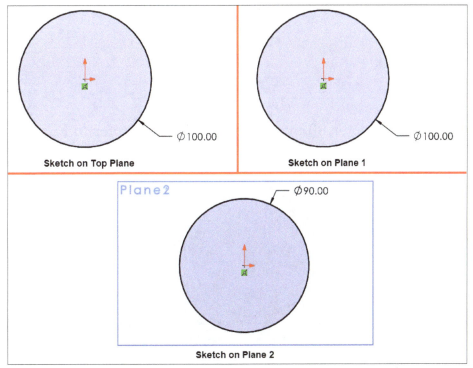

*Figure-4. Sketches created on various planes*

- To create 4th plane, you will need an axis as this plane is created at an angle of 5 degree with respect to horizontal line. Select the Plane 3 from graphics area and click on the **Sketch** tool from **Sketch CommandManager** in the **Ribbon**. The sketching environment will become active.
- Create a centerline as shown in Figure-5 and exit the sketch.

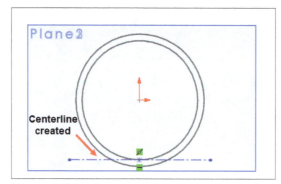

*Figure-5. Centerline created for plane*

- Click on the **Plane** tool from **Reference Geometry** drop-down in **Ribbon** and select the geometries as shown in Figure-6. Click on the **OK** button from **PropertyManager** to create the plane.

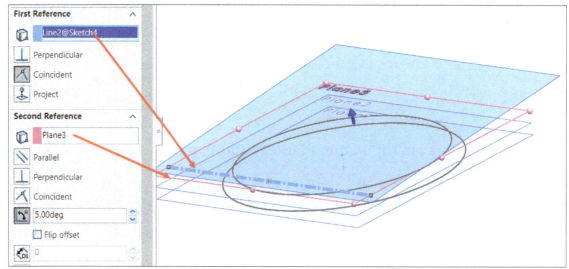

*Figure-6. Creating plane at an angle*

- Create a circle of diameter 80 mm on the new plane; refer to Figure-7.

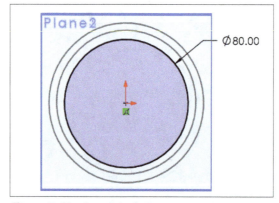

*Figure-7. Sketch on 4th plane*

## Creating Loft Feature

- Click on the **Lofted Surface** tool from the **Surfaces CommandManager** in the **Ribbon**. The **Surface-Loft PropertyManager** will be displayed.
- Select the sketches as shown in Figure-8. Preview of loft feature will be displayed.

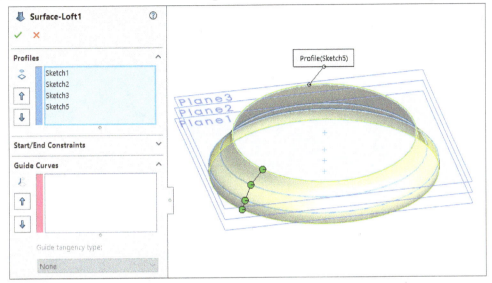

*Figure-8. Preview of surface loft*

- Click on the **OK** button from the **PropertyManager** to create the loft feature.

## Creating Fill Surface

- Click on the **Filled Surface** tool from **Surfaces CommandManager** in the **Ribbon**. The **Surface-Fill PropertyManager** will be displayed.
- Select the top edge of the model. Preview of fill surface will be displayed; refer to Figure-9.

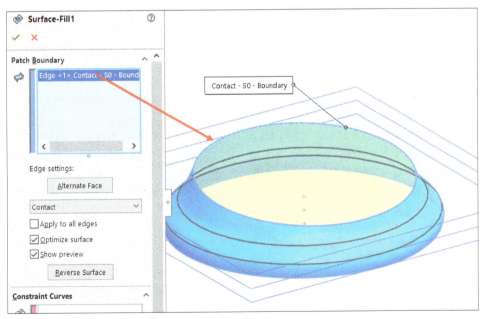

*Figure-9. Preview of fill surface*

- Set the parameters as shown in above figure and click on the **OK** button to create the surface.

## Creating Cut Feature

- Create a new plane at an offset distance of 20 mm above the Top plane; refer to Figure-10.

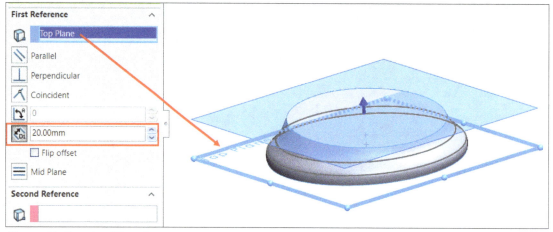

*Figure-10. Creating Plane 5*

- Select this newly created plane and create the sketch as shown in Figure-11.

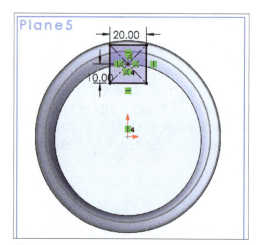

*Figure-11. Sketch created on Plane 5*

- Click on the **Planar Surface** tool from the **Surfaces CommandManager** in the **Ribbon**. The **Planar Surface PropertyManager** will be displayed.
- Select the sketch section from graphics area created earlier on Plane 5. Preview of planar surface will be displayed; refer to Figure-12. Click on the **OK** button from **PropertyManager** to create the surface.

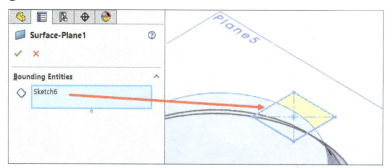

*Figure-12. Preview of planar surface*

- Click on the **Extruded Surface** tool from the **Surfaces CommandManager** in the **Ribbon**. The **Extrude PropertyManager** will be displayed.

- Select the sketch earlier used for planar surface and create extruded surface as shown in Figure-13. Click on the **OK** button from the **PropertyManager** to create surface.

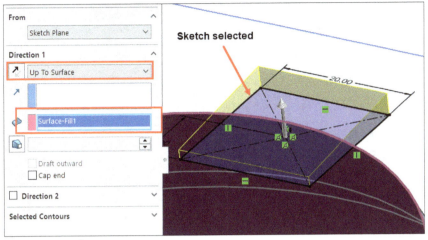

*Figure-13. Extruded surface*

- Now, you need to trim extra portion of surfaces to get a seamless surface. Click on the **Trim Surface** tool from the **Surfaces CommandManager** in the **Ribbon**. The **Trim Surface PropertyManager** will be displayed.
- Select the **Mutual** radio button from **PropertyManager** and set the parameters as shown in Figure-14. Click on the **OK** button from the **PropertyManager** to create the trimmed surface; refer to Figure-15.

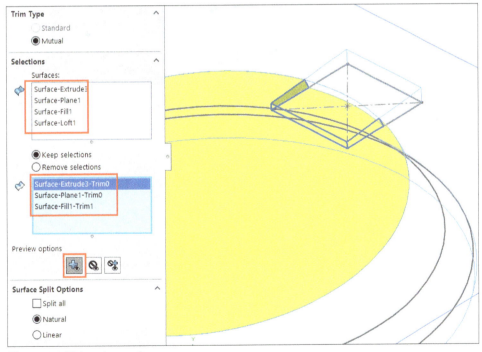

*Figure-14. Trimming surface*

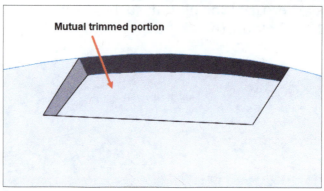

*Figure-15. Mutual trimmed surface*

- Click on the **Trim Surface** tool again and set the parameters as shown in Figure-16.

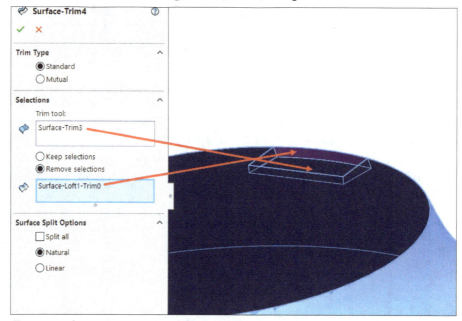

*Figure-16. Creating trimmed surface*

- Click on the **OK** button from **PropertyManager** to create trimmed surface; refer to Figure-17.

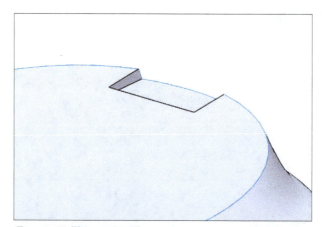

*Figure-17. Trimmed surface*

## Knitting Surfaces and Thickening

- Click on the **Knit Surface** tool from the **Surfaces CommandManager** in the **Ribbon**. The **Knit Surface PropertyManager** will be displayed.
- Select the surface from graphics area to be combined; refer to Figure-18.

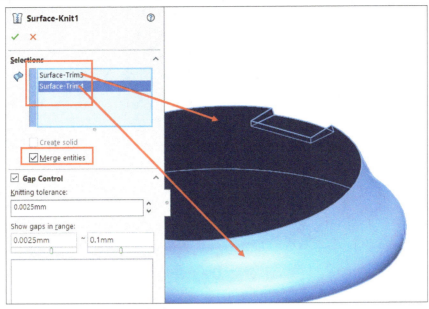

*Figure–18. Surfaces selected for knitting*

- Click on the **OK** button from the **PropertyManager** to create merged surface.
- Click on the **Thicken** tool from **Ribbon** and select the knitted surface.
- Set the thickness as **2** mm in the **Thickness** edit box and other parameters as shown in Figure-19. Click on the **OK** button to get final model.

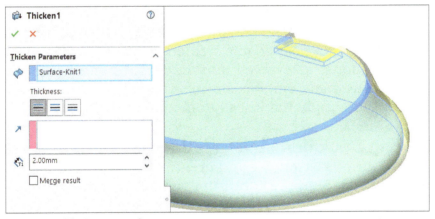

*Figure–19. Thickening surface*

## PRACTICAL 2 (MODERATE LEVEL)

Create the surface model using drawing shown in Figure-20 and set thickness of model as 2 mm.

*Figure-20. Surface Model Practical 2*

## Observation on Drawing / Blueprint Reading

The model does not have thickness so it is created using surface modeling. The model is created in three sections: a revolved surface feature, curved face created using trim and merge, and a hexagon emboss feature created by joining split surface. Knit all the surfaces and then apply thickness to model.

## Starting Document and Creating Revolved Surface

*   Start SolidWorks if not started yet and create a new Part document.
*   Click on the **Revolved Surface** tool from the **Surfaces CommandManager** in the **Ribbon**. The **Revolve PropertyManager** will be displayed.
*   Select the **Front Plane** from graphics area and create the sketch as shown in Figure-21. After creating sketch, click on the **Exit Sketch** button from **Ribbon**. Preview of surface revolve feature will be displayed.
*   Set the parameters as shown in Figure-22 and click on the **OK** button from **PropertyManager** to create the surface.

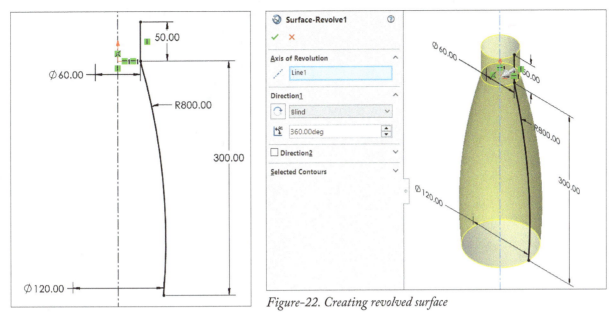

Figure-21. Sketch for revolved surface

Figure-22. Creating revolved surface

## Creating Curved Face

- Click on the **Extruded Surface** tool from the **Surfaces CommandManager** in the **Ribbon**. The **Extrude PropertyManager** will be displayed.
- Select Front Plane as sketch plane and create the sketch as shown in Figure-23.

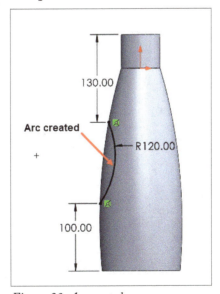

Figure-23. Arc created

- After creating sketch, click on the **Exit Sketch** tool from **Ribbon**. Preview of extruded surface will be displayed.
- Set the parameters as shown in Figure-24 and click on the **OK** button to create the surface.

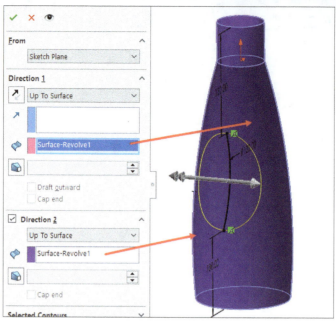

*Figure-24. Creating extruded surface*

- Click on the **Trim Surface** tool from the **Ribbon**. The **Trim Surface PropertyManager** will be displayed.
- Select the two surfaces and then set the parameters as shown in Figure-25. Click on the **OK** button from the **PropertyManager** to create the surface; refer to Figure-26.

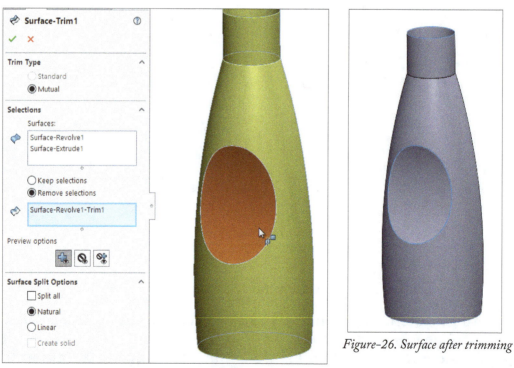

*Figure-26. Surface after trimming*

*Figure-25. Trimming surfaces*

## Creating Hexagon Emboss Feature

- Click on the Right Plane from graphics area and start a new sketch.
- Create a hexagon as shown in Figure-27 and click on the **Exit Sketch** button.

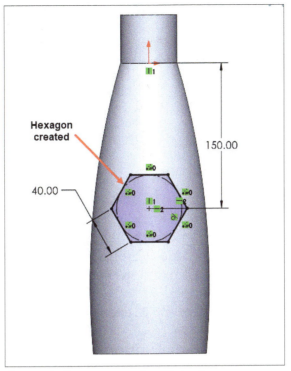

*Figure-27. Hexagon created on Right plane*

- Click on the **Split Line** tool from the **Curves** drop-down in the **Surfaces CommandManager** of the **Ribbon**. The **Split Line PropertyManager** will be displayed.
- Select the **Projection** radio button from the **PropertyManager** and select the hexagon sketch from graphics area. You will be asked to select surface for projection.
- Select the surface model and set parameters as shown in Figure-28.
- Click on the **OK** button from the **PropertyManager** to create split surface; refer to Figure-29.

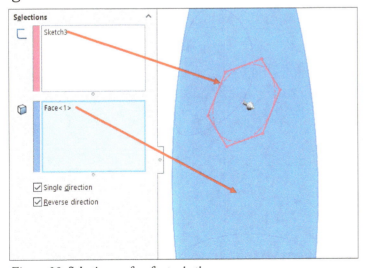

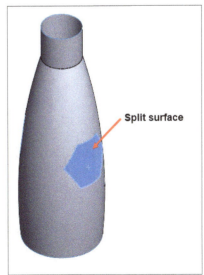

*Figure-28. Selecting surface for projection*

*Figure-29. Split surface created*

- Click on the **Offset Surface** tool from the **Surfaces CommandManager** in the **Ribbon**. The **Offset Surface PropertyManager** will be displayed.
- Select the split surface and specify offset distance as 5 mm; refer to Figure-30.

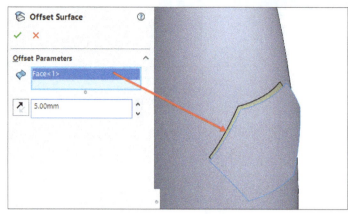

*Figure-30. Creating offset surface*

- Click on the **OK** button from the **PropertyManager** to create offset surface.
- Now, select the original split surface which was used for creating surface offset and click on the **Delete Face** tool from **Surfaces CommandManager** in the **Ribbon**. The selected surface will be deleted.
- Click on the **Ruled Surface** tool from the **Ribbon**. The **Ruled Surface PropertyManager** will be displayed.
- Select the **Normal to Surface** radio button from **Type** rollout to create surface perpendicular to base surface.
- Set the distance value as 5 mm in the **Distance/Direction** edit box of **PropertyManager**.
- Select the edges of deleted surface. Preview of ruled surface will be displayed; refer to Figure-31.

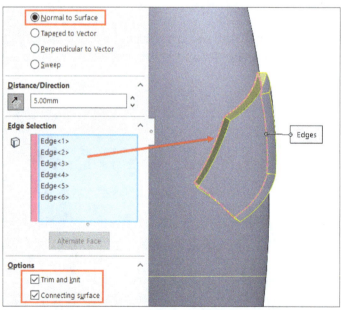

*Figure-31. Preview of ruled surface*

- Click on the **OK** button from the **PropertyManager** to create the surface.

## Creating Filled Surface and Applying Thickness

- Click on the **Filled Surface** tool from the **Surfaces CommandManager** in the **Ribbon**. The **Fill Surface PropertyManager** will be displayed.
- Select the bottom edge of the model and set parameters as shown in Figure-32.

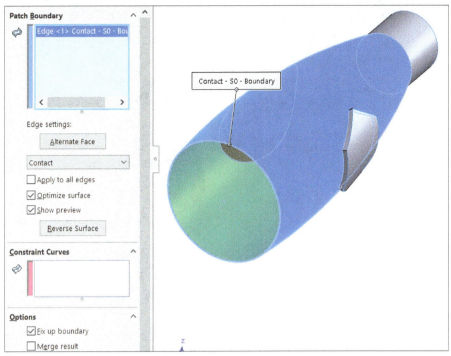

*Figure-32. Preview of filled surface*

- Click on the **OK** button to create the surface.
- Click on the **Knit Surface** tool from the **Ribbon**. The **Knit Surface PropertyManager** will be displayed.
- Select all the surfaces of the model to knit them together and set the parameters as shown in Figure-33.

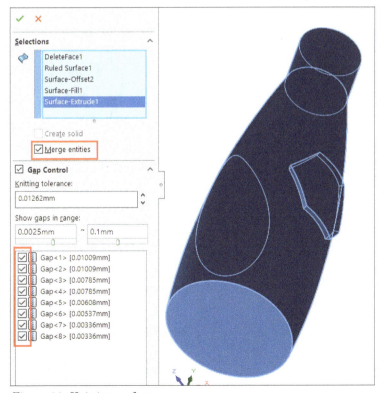

*Figure-33. Knitting surfaces*

- Click on the **OK** button from the **PropertyManager** to create the surface.
- Click on the **Thicken** tool from the **Ribbon**. The **Thicken PropertyManager** will be displayed.

• Select the model from graphics area and set the parameters as shown in Figure-34.

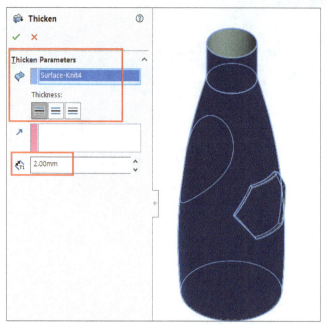

*Figure-34. Applying thickness to model*

• Click on the **OK** button from the **PropertyManager** to create the final model.

## PRACTICAL 3 (ADVANCED SURFACE DESIGN)

Create the model of helmet shell as shown in Figure-35.

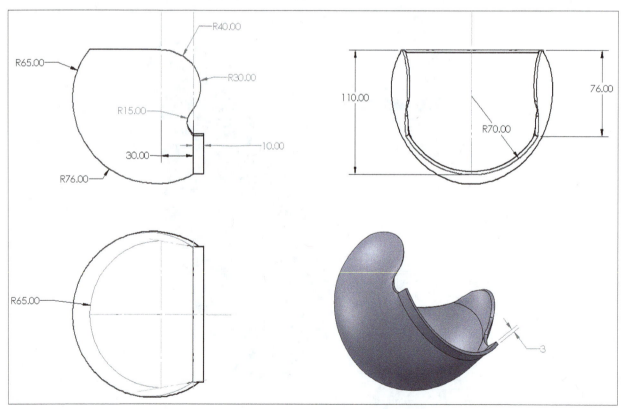

*Figure-35. Surface Model Practical 3*

## Observation on Drawing / Blueprint Reading

The model comprises of complex shape following various curves so it is better to use surface modeling. The model can be created using loft surface feature which will use open section curves with various guide curves. There is a small section of surface extended from the top of helmet shell which can be created using ruled surface feature. After that knit the surface and apply thickness to it.

## Starting Model and Creating Sketches

- Start SolidWorks if not started yet.
- Click on the **New** tool from **File** menu or **Quick Access Toolbar**. The **New SOLIDWORKS Document** dialog box will be displayed.
- Double-click on **Part** button from the dialog box. A new part document will open.
- Click on the **Plane** tool from **Reference Geometry** drop-down in **Features CommandManager** in the **Ribbon**. The **Plane PropertyManager** will be displayed.
- Select the **Right Plane** from graphics area and create a plane at a distance of 30 mm; refer to Figure-36.

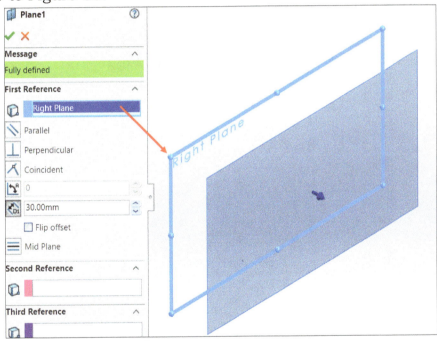

*Figure-36. Creating offset plane1*

- Click on the **OK** button to create the plane.
- Select the Front Plane from graphics area and create an arc in sketch as shown in Figure-37.

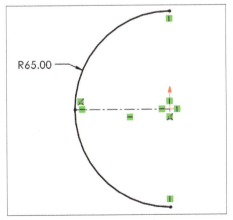

*Figure-37. Sketch on Front plane*

- Select the Plane 1 recently created and create the sketch as shown in Figure-38.

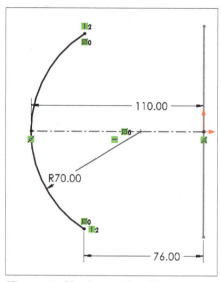

*Figure-38. Sketch created on Plane1*

- Now, we will create guide curves for loft surface feature. Select the **Top Plane** from graphics area and create the sketch as shown in Figure-39.

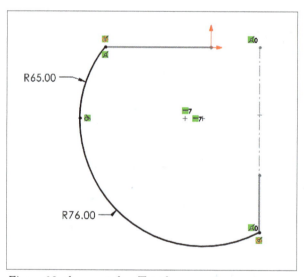

*Figure-39. Arcs created on Top plane*

- Click on the **Plane** tool from the **Reference Geometry** drop-down in the **Features CommandManager** in the **Ribbon**. The **Plane PropertyManager** will be displayed.
- Create a plane passing through end points of section curves and perpendicular to Front plane; refer to Figure-40.

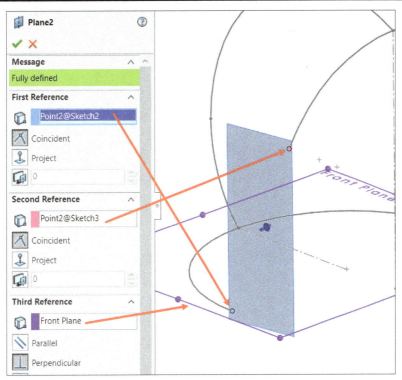

*Figure-40. Creating Plane 2*

- After creating plane, create the sketch of side profile as shown in Figure-41 on the plane.

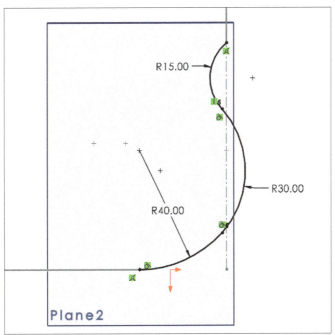

*Figure-41. Creating sketch for side profile*

- Create similar plane and sketch on other end points of section curves; refer to Figure-42.

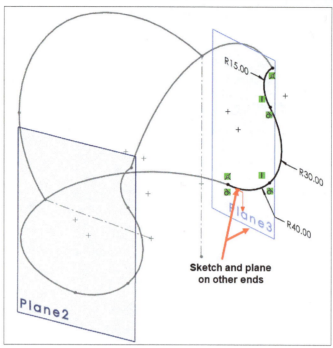

*Figure-42. Sketch on other end points*

• Create a new plane passing through three points of guide curves as shown in Figure-43.

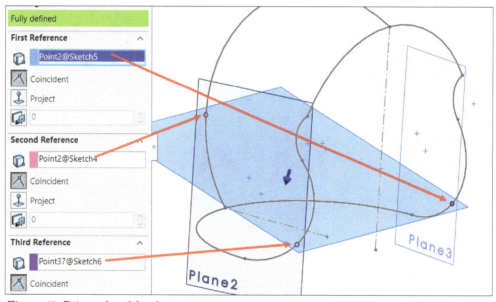

*Figure-43. Points selected for plane*

• Create the sketch as shown in Figure-44 on the newly created plane.

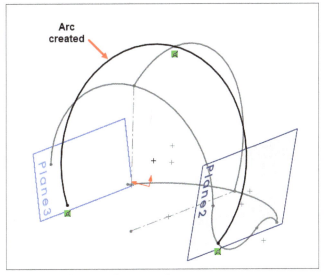

*Figure-44. Arc created on new plane*

## Creating Loft Surface Feature

- Click on the **Lofted Surface** tool from the **Surfaces CommandManager** in the Ribbon. The **Surface-Loft PropertyManager** will be displayed.
- Select the section curves and guide curves as shown in Figure-45.

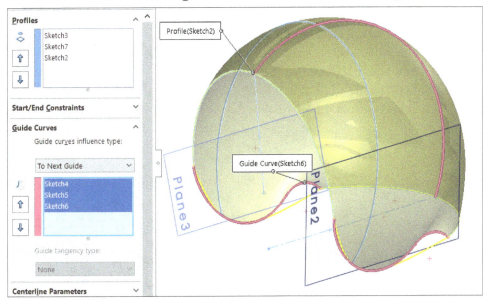

*Figure-45. Preview of loft surface*

- Click on the **OK** button from **PropertyManager** to create the surface.

## Creating Ruled Surface

- Click on the **Ruled Surface** tool from the **Surfaces CommandManager** in the **Ribbon**. The **Ruled Surface PropertyManager** will be displayed.
- Select the top edge of surface model and set the parameters as shown in Figure-46.

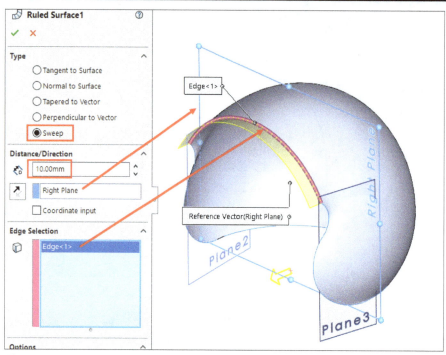

*Figure-46. Creating ruled surface*

- Click on the **OK** button from **PropertyManager** to create the extended surface.

Tip: You might ask why not to use **Extend Surface** tool in place of **Ruled Surface** tool to create extended surface. Note that if you use **Extend Surface** tool then new surface will follow the curvature of base surface but if you use **Ruled Surface** tool then you can get linear extension of the surface as in our case. You can check difference between two extended surfaces in Figure-47.

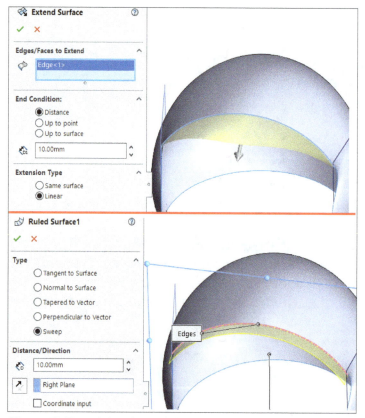

*Figure-47. Extended surface vs ruled surface*

## Applying Thickness and Cleaning Model

- Click on the **Knit Surface** tool from the **Surfaces CommandManager** in the **Ribbon**. The **Surface-Knit PropertyManager** will be displayed.
- Select the loft surface and ruled surface from graphics area and set the parameters as shown in Figure-48.

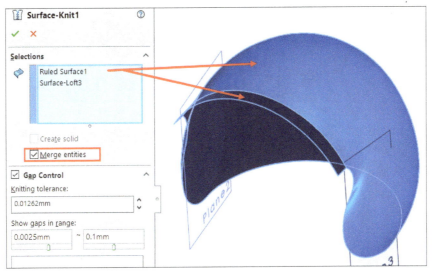

*Figure-48. Knitting loft and ruled surfaces*

- Click on the **OK** button from **PropertyManager** to create knitted surface.
- Click on the **Thicken** tool from the **Ribbon** and select the knitted surface. Preview of feature will be displayed.
- Set the thickness direction inwards and thickness value as 3 mm; refer to Figure-49.

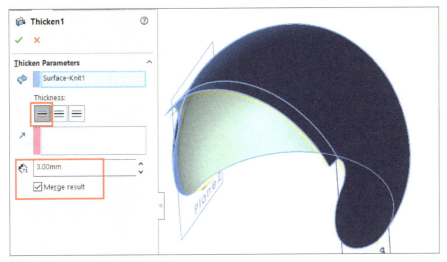

*Figure-49. Thicken parameters*

- Click on the **OK** button from the **PropertyManager** to create solid body. Note that a small portion of material is created at the bottom of body as shown in Figure-50. You can use extrude cut feature to remove this extra material.

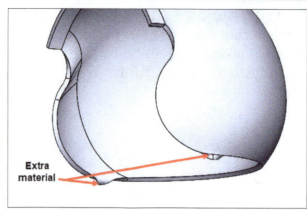

*Figure-50. Extra material in model*

- Click on the **Extruded Cut** tool from the **Features CommandManager** in the **Ribbon**. The **Extrude PropertyManager** will be displayed and you will be asked to select a sketching plane.
- Select the Top Plane from graphics area and create the sketch as shown in Figure-51.

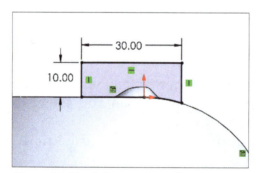

*Figure-51. Sketch for extrude cut feature*

- Click on the **Exit Sketch** tool from graphics area or **Ribbon**. Preview of extrude cut feature will be displayed.
- Select the **Through All - Both** option from **End Condition** drop-down in **Direction 1** rollout of **PropertyManager**; refer to Figure-52.

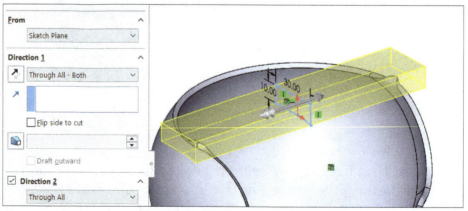

*Figure-52. Preview of extrude cut*

- Click on the **OK** button to create the feature.

## PRACTICE 1 TO 4

Create the models given in next figures.

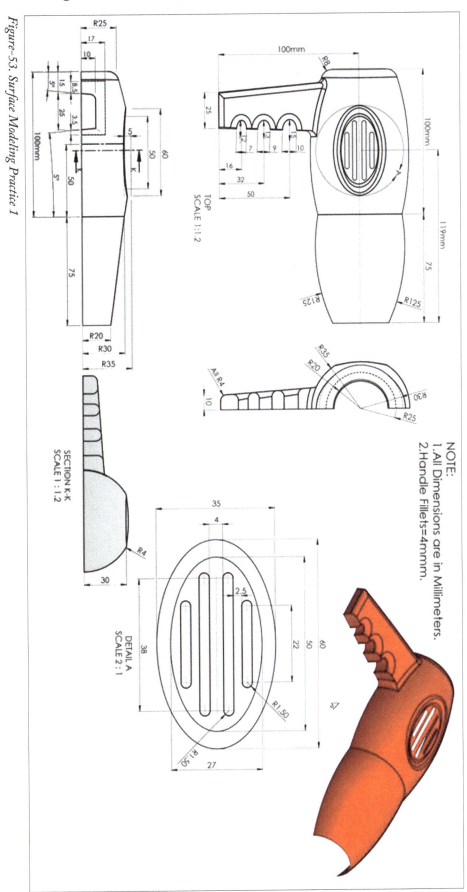

*Figure-53. Surface Modeling Practice 1*

NOTE:
1. All Dimensions are in Millimeters.
2. Handle Fillets=4mmm.

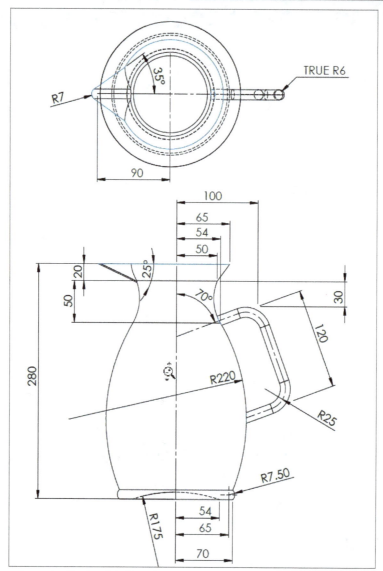

*Figure-54. Surface Modeling Practice 2*

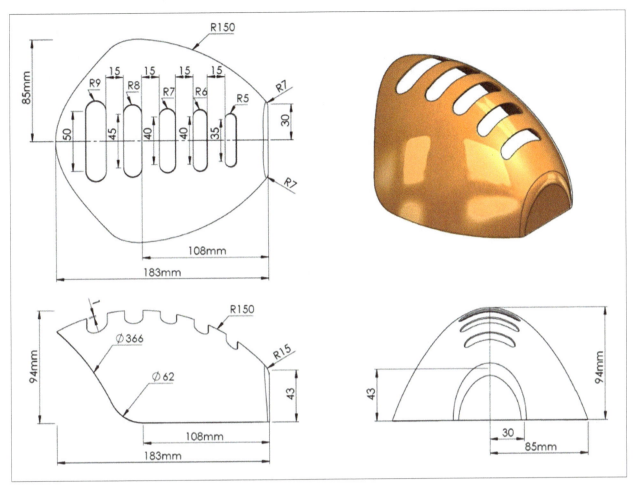

*Figure-55. Surface Modeling Practice 3*

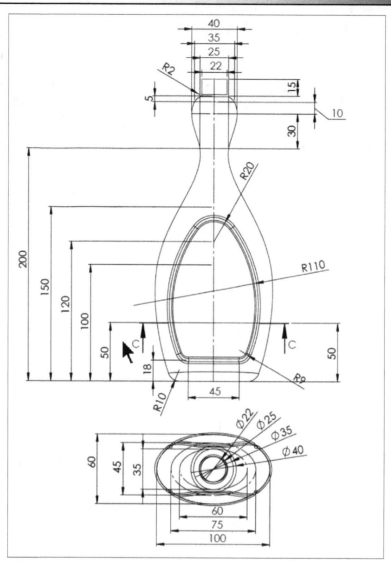

*Figure-56. Surface Modeling Practice 4*

# FOR STUDENT NOTES

# FOR STUDENT NOTES

# Chapter 5

# Sheetmetal Design
# Practical and Practice

## Topics Covered

The major topics covered in this chapter are:

- *Basic Sheetmetal Design*
- *Advanced Sheetmetal Design*

# INTRODUCTION

Sheetmetal design is used when thickness of model is less than 2 inches and it has bends of different angles manufactured by press machines. Sheetmetal models also have cuts made by punching tools. The purpose of using CAD for sheetmetal design is to find out flat pattern of sheetmetal model with detailed bend table. Flat pattern provides the shape and size of sheet metal piece to be used for manufacturing model. The bend table provides data on where a bend will be applied, what will be bend angle and what will be the direction of bend. The bends are marked as bend lines on the flat pattern. In this chapter, you will learn to create various types of sheetmetal models.

# PRACTICAL 1 (BASIC LEVEL)

Create the model as shown in Figure-1.

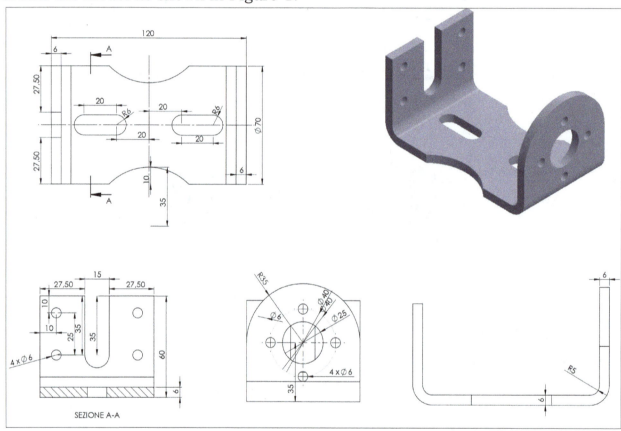

*Figure-1. Sheetmetal Design Practical 1*

## Observation on Drawing / Blueprint Reading

The thickness of model is 6 mm and model contains bends and cuts only. So, this model can be created by sheetmetal designing methods. In this model, there is a base created using **Base Flange/Tab** tool. The walls of the model are created using the **Edge Flange** tool. The steps to create model are given next.

## Starting New Part Document

- Start SolidWorks if not started yet.
- Click on **New** tool from **File** menu or press **CTRL+N** using Keyboard. The **New SOLIDWORKS Document** dialog box will be displayed.
- Double-click on **Part** button from the dialog box. A new part document will open.
- Click on the **Sheet Metal CommandManager** from the **Ribbon** to display tools for sheet metal design.

- If the **Sheet Metal** tab is not available in **Ribbon** by default then right-click on any tool of the **Ribbon** and select the **Sheet Metal** option from **Tabs** cascading menu of right-click shortcut menu; refer to Figure-2.

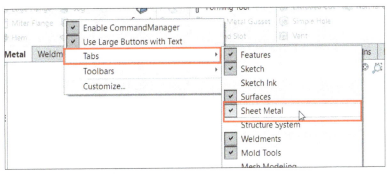

*Figure-2. Sheet Metal option*

## Creating Base Flange/Tab

- Click on the **Base Flange/Tab** tool from the **Sheet Metal CommandManager** in the **Ribbon**. You will be asked to select a plane for creating base sketch.
- Select the **Top Plane** from graphics area. The sketching environment will become active.
- Create sketch as shown in Figure-3 and click on the **Exit Sketch** tool from **Ribbon**. Preview of base flange will be displayed.

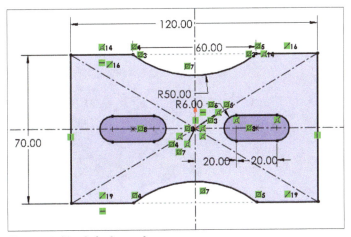

*Figure-3. Sketch for base tab*

- Set the thickness of base flange as **6** mm in the **Direction 1 Thickness** edit box of the **Sheet Metal Parameters** rollout in the **Base Flange PropertyManager** and click on the **OK** button to create the base feature; refer to Figure-4.

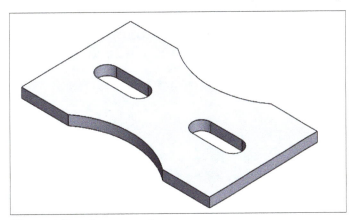

*Figure-4. Base flange created*

## Creating Edge Flange

- Click on the **Edge Flange** tool from the **Ribbon**. The **Edge Flange PropertyManager** will be displayed.
- Select the bottom side edge of model. Preview of edge flange will be displayed with end point of flange attached to cursor; refer to Figure-5.

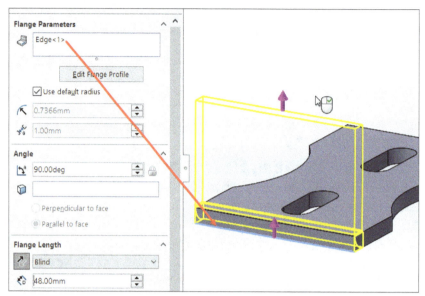

*Figure-5. Preview of edge flange*

- Move the cursor upward and click when height of flange is approximately 35 mm; refer to Figure-6.

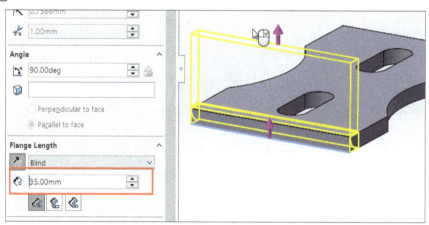

*Figure-6. Setting height of flange*

- Clear the **Use default radius** check box and specify the radius as **5** mm in the **Bend Radius** edit box.
- Click on the **Edit Flange Profile** button from the **Flange Parameters** rollout in **PropertyManager**. The profile sketch of flange will be displayed and sketching tools will become active; refer to Figure-7.

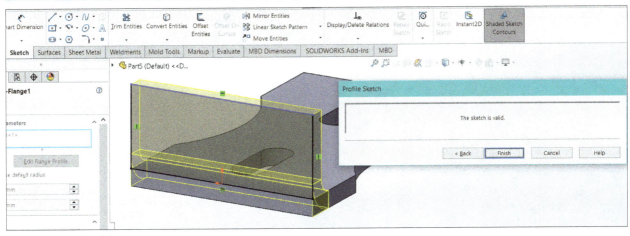

*Figure-7. Profile of flange*

- Modify the profile sketch as shown in Figure-8. After creating profile, click on the **Finish** button from **Profile Sketch** dialog box. The model will be displayed as shown in Figure-9.

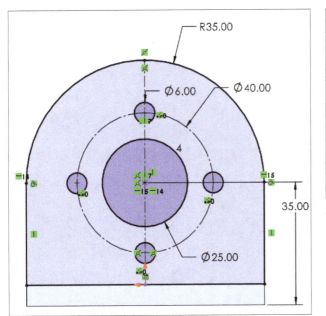

*Figure-8. Profile of edge flange*

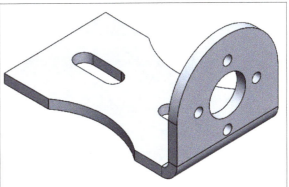

*Figure-9. Model after creating edge flange*

- Similarly, create edge flange on other side with profile sketch as shown in Figure-10. The edge flange preview will display as shown in Figure-11.

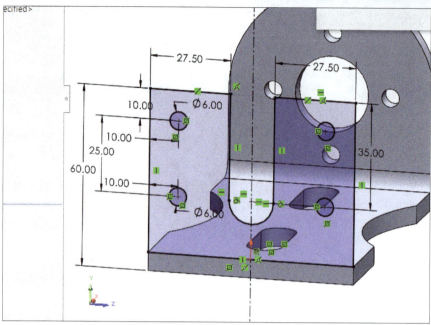

*Figure-10. Edge flange profile on opposite edge*

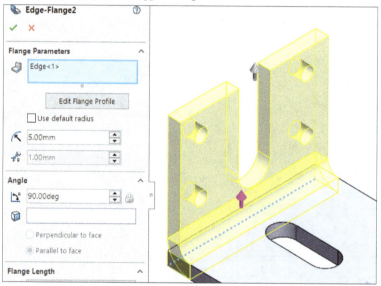

*Figure-11. Edge flange on opposite edge of model*

- Click on the **OK** button from **PropertyManager** to create the flange; refer to Figure-12.

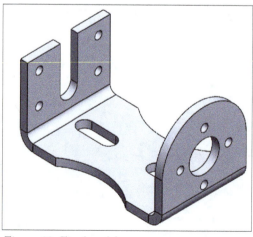

*Figure-12. Final model*

# PRACTICAL 2 (MODERATE LEVEL)

Create the sheetmetal model using drawing shown in Figure-13.

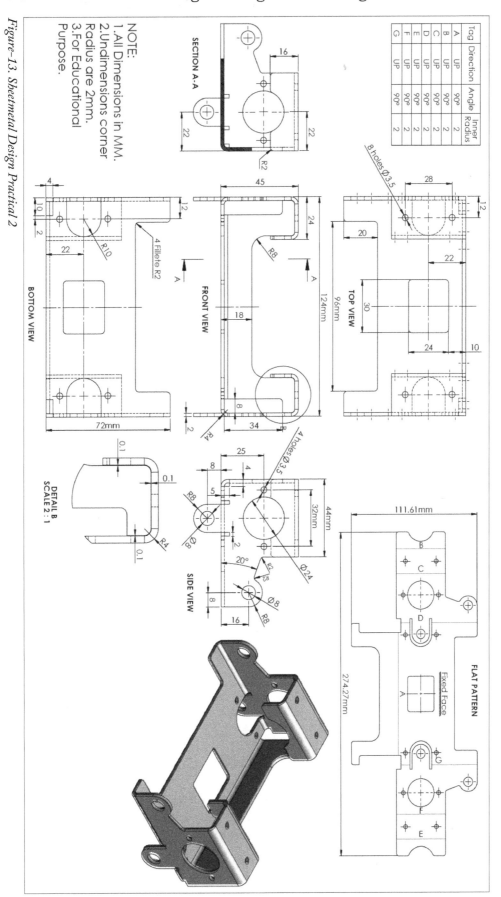

| Tag | Direction | Angle | Inner Radius |
|---|---|---|---|
| A | UP | 90° | 2 |
| B | UP | 90° | 2 |
| C | UP | 90° | 2 |
| D | UP | 90° | 2 |
| E | UP | 90° | 2 |
| F | UP | 90° | 2 |
| G | UP | 90° | 2 |

NOTE:
1. All Dimensions in MM.
2. Undimensions corner Radius are 2mm.
3. For Educational Purpose.

*Figure-13. Sheetmetal Design Practical 2*

## Observation on Drawing / Blueprint Reading

The model is a complex sheetmetal part which can be created by combining multiple features. These features are: Base feature created using **Base Flange/Tab** tool, one side flange created using combination of **Edge Flange**, **Unfold**, **Extrude cut**, **Base Flange/Tab**, and then **Fold** tool, other side flange created by mirroring, and rest of the flanges created using the **Edge Flange** tool. The steps to create model are given next.

## Starting a New Document

- Start SolidWorks if not started yet.
- Click on **New** tool from **File** menu or press **CTRL+N** using Keyboard. The **New SOLIDWORKS Document** dialog box will be displayed.
- Double-click on **Part** button from the dialog box. A new part document will open.
- Click on the **Sheet Metal CommandManager** from the **Ribbon** to display tools for sheet metal design.

## Creating Base Tab

- Click on the **Base Flange/Tab** tool from the **Sheet Metal CommandManager** in the **Ribbon**. You will be asked to select a plane for creating profile sketch.
- Select the Top Plane from graphics area. The sketching environment will become active.
- Create the sketch as shown in Figure-14.

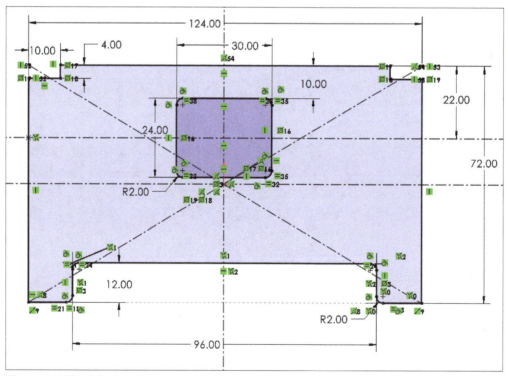

*Figure-14. Sketch for base flange*

- Click on the **Exit Sketch** tool from the **Ribbon**. The **Base Flange PropertyManager** will be displayed with preview of base flange.
- Specify the thickness of sheet as **2 mm** in the **Thickness** edit box of **Sheet Metal Parameters** rollout in the **PropertyManager**. Keep the other parameters as default and click on the **OK** button from **PropertyManager**. The base flange will be displayed as shown in Figure-15.

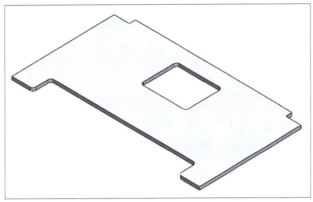

*Figure-15. Base flange*

## Creating Side Wall Flange

- Click on the **Edge Flange** tool from the **Sheet Metal CommandManager** in the **Ribbon**. The **Edge Flange PropertyManager** will be displayed and you will be asked to select an edge.
- Select the left side edge of model and set parameters as shown in Figure-16.

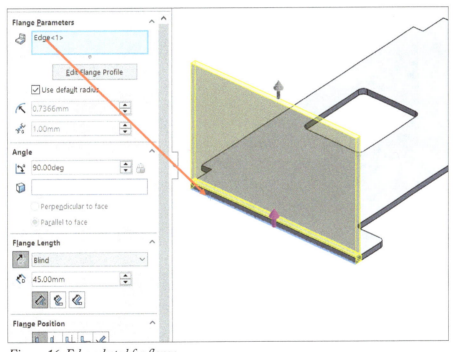

*Figure-16. Edge selected for flange*

- Click on the **Edit Flange Profile** button from the **PropertyManager** and create the profile as shown in Figure-17.

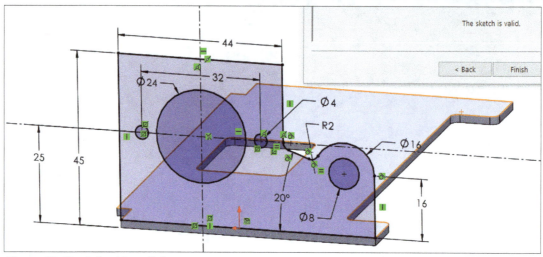

*Figure-17. Sketch for side wall flange*

- Click on the **Finish** button from the **Profile Sketch** dialog box. The edge flange will be created; refer to Figure-18.

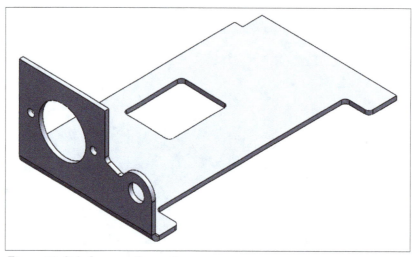

*Figure-18. Side flange wall created*

## Creating Extrude Cut and Tab at Bend

- Click on the **Unfold** tool from the **Sheet Metal CommandManager** in the **Ribbon**. The **Unfold PropertyManager** will be displayed.
- Select the fixed face and bend to be unfold from the model as shown in Figure-19.

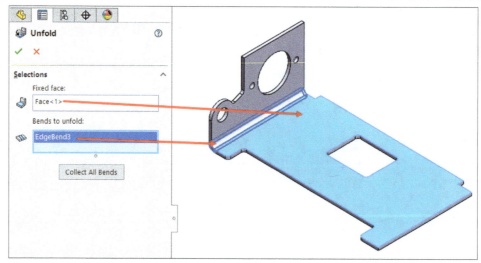

*Figure-19. Unfold PropertyManager*

- Click on the **Extruded Cut** tool from the **Sheet Metal CommandManager** in the **Ribbon**. You will be asked to select a plane/face for creating sketch.
- Select the top face of unfolded sheet metal model. The sketching environment will become active.
- Create the sketch as shown in Figure-20 and then click on the **Exit Sketch** tool from **Ribbon**. Preview of extruded cut feature will be displayed; refer to Figure-21.

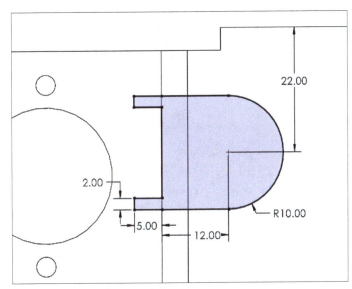

*Figure-20. Sketch for extruded cut*

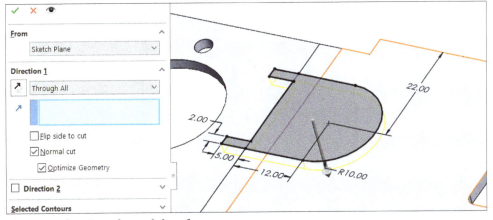

*Figure-21. Preview of extruded cut feature*

- Select the **Through All** option from the **End Condition** drop-down in **Direction 1** rollout of **PropertyManager** and click on the **OK** button from **PropertyManager**. The cut will be created; refer to Figure-22.

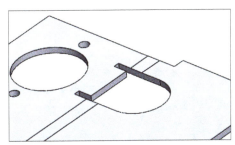

*Figure-22. Extruded cut created*

- Click on the **Fold** tool from the **Ribbon**. The **Fold PropertyManager** will be displayed.
- Set the parameters as shown in Figure-23 and click on the **OK** button to fold the bend again.

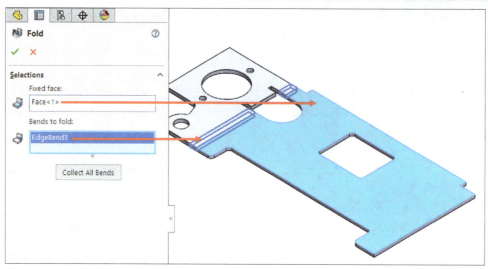

*Figure-23. Parameters specified in Fold PropertyManager*

- Click on the **Base Flange/Tab** tool from the **Sheet Metal CommandManager** in the **Ribbon**. You will be asked to select a planar face or plane.
- Select the planar face of side wall as shown in Figure-24. The sketching environment will become active.

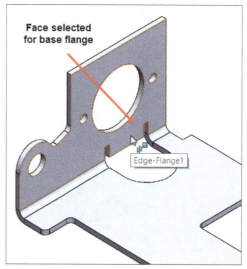

*Figure-24. Planar face selected*

- Create the sketch as shown in Figure-25 and then click on the **Exit Sketch** tool from the **Ribbon**. Preview of flange will be displayed; refer to Figure-26.

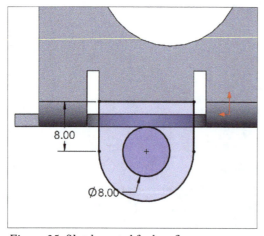

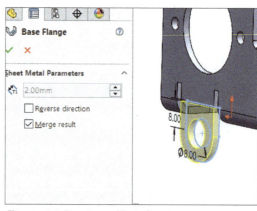

*Figure-26. Preview of base flange*

*Figure-25. Sketch created for base flange*

- Click on the **OK** button from **PropertyManager** to create the base flange.

## Creating Mirror Copy

When creating mirror copy of sheetmetal features in SolidWorks, you cannot create mirror copy of native sheetmetal features like Unfold, Fold, and so on. So, you need to create these features manually while creating mirror copies of other sheetmetal features.

- Select the **Edge-Flange1** feature from the **FeatureManager Design Tree** and click on the **Mirror** tool from **Features CommandManager** in the **Ribbon**. The **Mirror PropertyManager** will be displayed and you will be asked to select mirror plane.
- Select the **Right Plane** from in-graphics **FeatureManager Design Tree**. Preview of mirror copy will be displayed; refer to Figure-27.

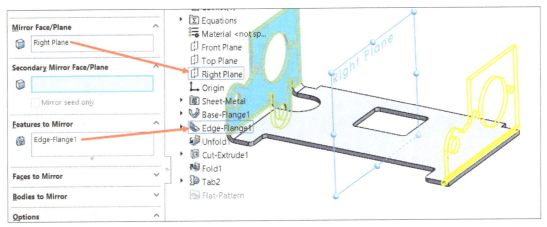

*Figure-27. Preview of mirroring*

- Click on the **OK** button from the **PropertyManager** to create the mirror copy; refer to Figure-28.

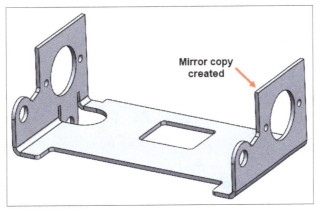

*Figure-28. Mirror copy of edge flange*

- Now, you need to unfold this mirror copy of flange manually as it cannot be mirror copied. Click on the **Unfold** tool from the **Sheet Metal CommandManager** in the **Ribbon**. The **Unfold PropertyManager** will be displayed.
- Select the flat face earlier selected as fixed face and select the bend of mirror copied flange; refer to Figure-29. Click on the **OK** button from **PropertyManager** to unfold the bend.

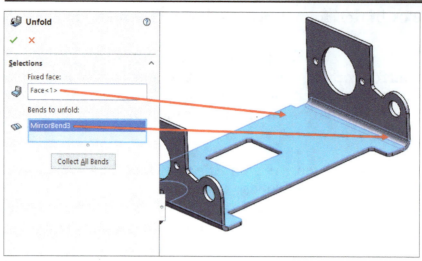

*Figure-29. Unfolding mirror copy*

- Select the **Cut-Extrude1** feature from **FeatureManager Design Tree** and click on the **Mirror** tool from **Features CommandManager** in the **Ribbon**. The **Mirror PropertyManager** will be displayed and you will be asked to select mirror plane.

- Select the **Right Plane** from in-graphics **FeatureManager Design Tree**. Preview of mirror copy will be displayed. Click on the **OK** button from **PropertyManager** to create copy of cut feature; refer to Figure-30.

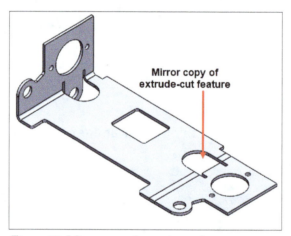

*Figure-30. Mirror copy of extrude-cut feature*

- Now, you need to fold the bend manually so click on the **Fold** tool from **Sheet Metal CommandManager** in the **Ribbon** and refold the bend earlier unfolded; refer to Figure-31.

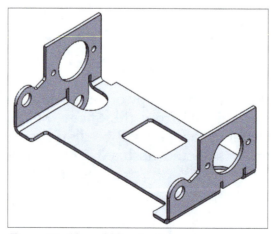

*Figure-31. After refolding bend*

- Select the **Tab1** feature from **FeatureManager Design Tree** and create its mirror copy with respect to **Right Plane**; refer to Figure-32.

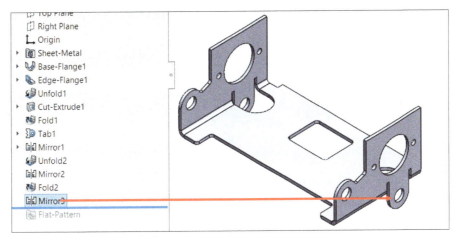

*Figure-32. Mirror copy of tab*

## Creating Remaining Flanges

- Click on the **Edge Flange** tool from the **Sheet Metal CommandManager** in the **Ribbon**. The **Edge Flange PropertyManager** will be displayed.
- Clear the **Use default radius** check box and specify bend radius as 2 mm.
- Set the length of flange as 24 mm. Preview of edge flange will be displayed; refer to Figure-33.

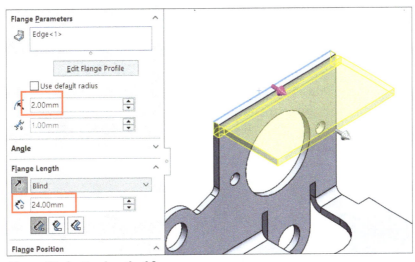

*Figure-33. Setting length of flange*

- Click on the **OK** button to create the flange.
- Click again on **Edge Flange** tool from the **Sheet Metal CommandManager** in the **Ribbon** and set the parameters as shown in Figure-34.
- Click on the **Edit Flange Profile** button from the **PropertyManager** and create profile sketch as shown in Figure-35.

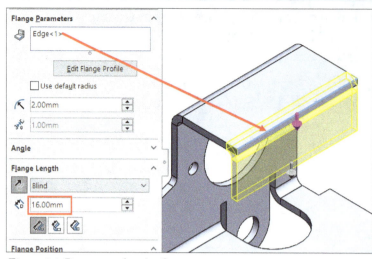

*Figure-34. Parameters for edge flange*

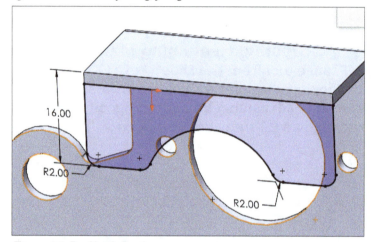

*Figure-35. Profile of edge flange*

- Click on the **Finish** button from the **Profile Sketch** dialog box.
- Create holes in the flange using the **Extruded Cut** tool; refer to Figure-36.

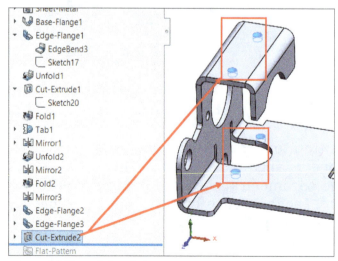

*Figure-36. Extruded cut holes*

- Select the last two edge flanges and extruded cut features from **FeatureManager Design Tree** and then click on the **Mirror** tool from the **Features CommandManager** in the **Ribbon**. You will be asked to select mirror plane.
- Select the **Right Plane** from **FeatureManager Design Tree**. Preview of mirror copy will be displayed; refer to Figure-37.

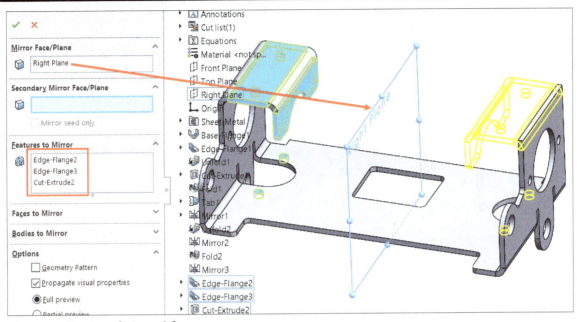

*Figure-37. Preview of mirrored flanges*

- Click on the **OK** button from **PropertyManager** to create the mirror copy of flanges.
- Click on the **Edge Flange** tool from the **Sheet Metal CommandManager** in the **Ribbon** and set the parameters as shown in Figure-38.

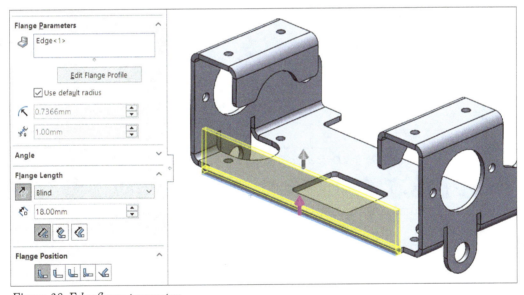

*Figure-38. Edge flange parameters*

- Click on the **Edit Flange Profile** button from **PropertyManager** and modify the sketch as shown in Figure-39.

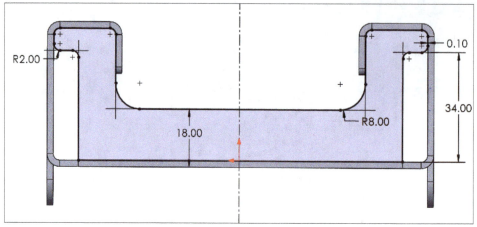

*Figure-39. Sketch for edge flange*

- Click on the **Finish** button from the **Profile Sketch** dialog box. The flange will be created; refer to Figure-40.

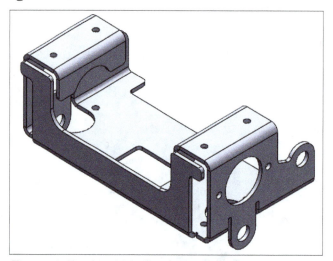

*Figure-40. Final model for Practical 2*

## PRACTICAL 3 (ADVANCED LEVEL)

Create the sheet metal model as shown in Figure-41.

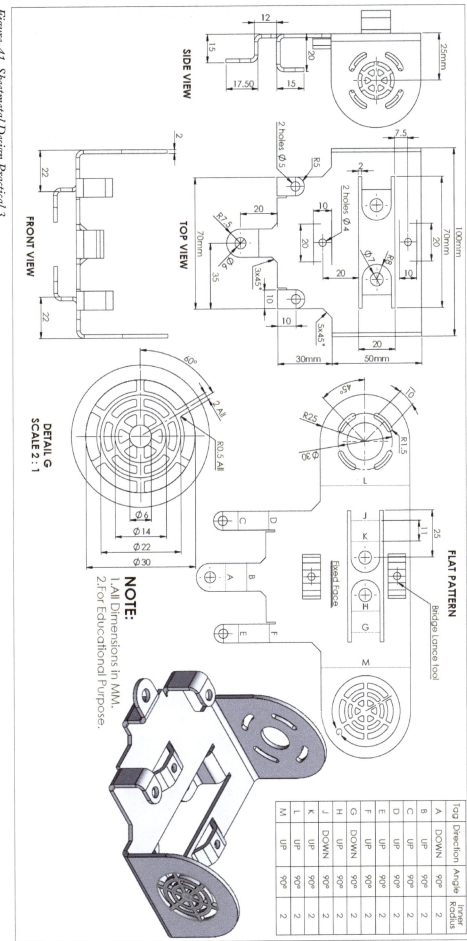

Figure-41. Sheetmetal Design Practical 3

## Observation on Drawing / Blueprint Reading

The model is a complex sheetmetal part which can be created by combining multiple features. These features are: Base feature created using **Base Flange/Tab** tool, one side flange created using combination of **Edge Flange, Unfold, Extrude cut, Base Flange/Tab,** and then **Fold** tool, other side flange created by mirroring, and rest of the flanges created using the **Edge Flange** tool. The steps to create model are given next.

## Starting a New Document

• Start SolidWorks if not started yet.
• Click on **New** tool from **File** menu or press **CTRL+N** using Keyboard. The **New SOLIDWORKS Document** dialog box will be displayed.
• Double-click on **Part** button from the dialog box. A new part document will open.
• Click on the **Sheet Metal CommandManager** from the **Ribbon** to display tools for sheet metal design.

## Creating Base Tab

• Click on the **Base Flange/Tab** tool from the **Sheet Metal CommandManager** in the **Ribbon**. You will be asked to select a plane for creating profile sketch.
• Select the **Top Plane** from graphics area. The sketching environment will become active.
• Create the sketch as shown in Figure-42 and click on the **Exit Sketch** button from the **Ribbon**. Preview of base flange will be displayed; refer to Figure-43.

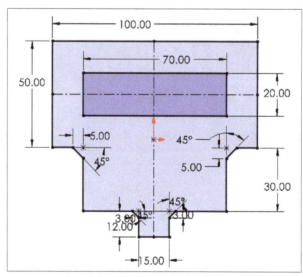

*Figure-42. Sketch for base flange feature*

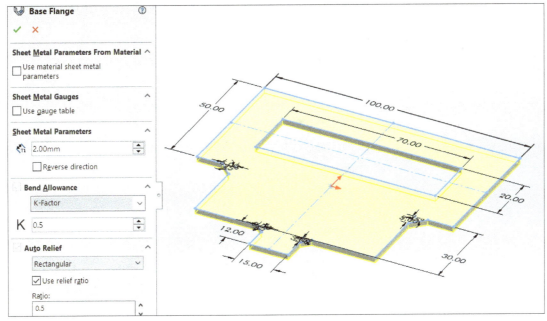

*Figure-43. Base flange preview*

- Click on the **OK** button from the **PropertyManager** to create flange.

## Creating Left Flange Wall

- Click on the **Edge Flange** tool from **Sheet Metal CommandManager** in the **Ribbon**. You will be asked to select an edge for creating flange.
- Select left edge of the model. Preview of edge flange will be displayed with end point of flange attached to cursor.
- Set the length of flange as 25 mm, bend radius as 2 mm, and keep the other parameters to default; refer to Figure-44.

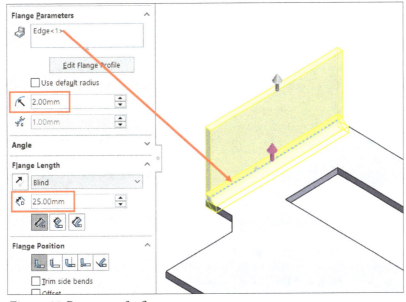

*Figure-44. Parameters for flange*

- Click on the **Edit Flange Profile** button from **PropertyManager**. The sketching environment will become active.
- Create the sketch as shown in Figure-45.

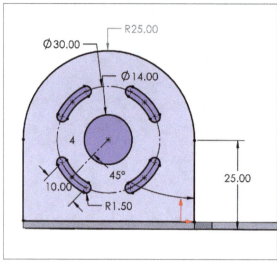

*Figure–45. Sketch for flange*

- Click on the **Finish** button from the **Profile Sketch** dialog box. The flange will be created.

## Creating Right Flange Wall

- Click on the **Edge Flange** tool from the **Sheet Metal CommandManager** in the **Ribbon**. The **Edge Flange PropertyManager** will be displayed and you will be asked to select edges.
- Select the right edge of the model. Preview of flange will be displayed with end point attached to cursor.
- Set the parameters as shown in Figure-46.

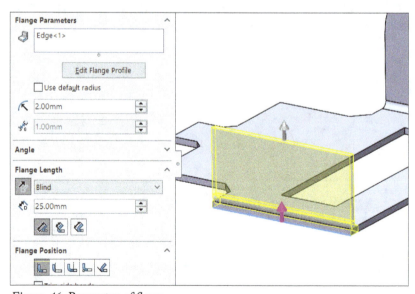

*Figure–46. Parameters of flange*

- Click on the **Edit Flange Profile** button and create the sketch profile as shown in Figure-47.

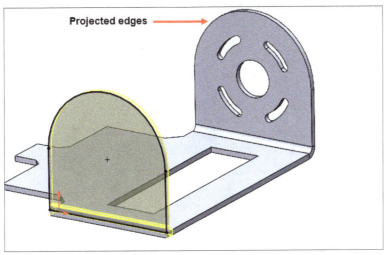

*Figure–47. Projected edges of flange*

- Click on the **Finish** button from the **Profile Sketch** dialog box to create the flange.
- Click on the **Sketch** tool from the **Sketch CommandManager** in the **Ribbon**. You will be asked to select a plane/planar face for creating sketch.
- Select the outer flat face of newly created flange and create the sketch as shown in Figure-48.

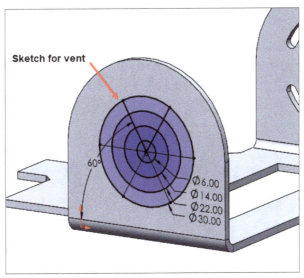

*Figure–48. Sketch for vent*

- After creating sketch, click on the **Vent** tool from the **Sheet Metal CommandManager** in the **Ribbon**. The **Vent PropertyManager** will be displayed.
- Select the entities and set parameters as shown in Figure-49.

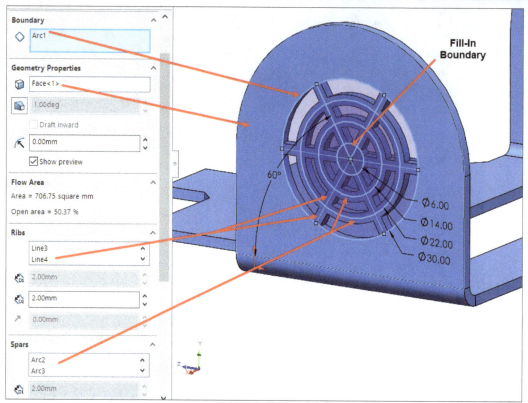

*Figure-49. Objects selected for vent*

- Click on **OK** button from the **PropertyManager** to create vent; refer to Figure-50.

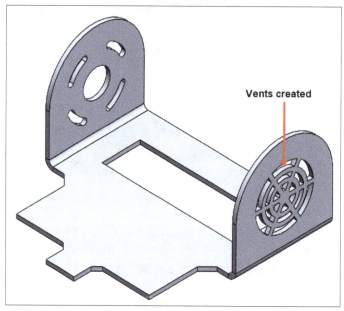

*Figure-50. Vents created*

- Click on the **Edge Flange** tool from the **Ribbon** and select the edge as shown in Figure-51. Click on the **OK** button from the **PropertyManager** to create the flange.

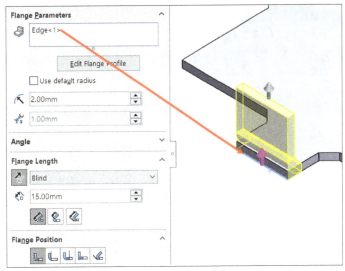

*Figure-51. Edge selected*

- Click again on the **Edge Flange** tool from the **Ribbon** and create the flange as shown in Figure-52 on edge of earlier created flange.

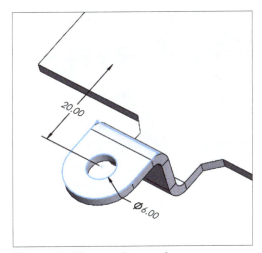

*Figure-52. Flange to be created*

- Similarly, create the other flanges on sides of this flange; refer to Figure-53.

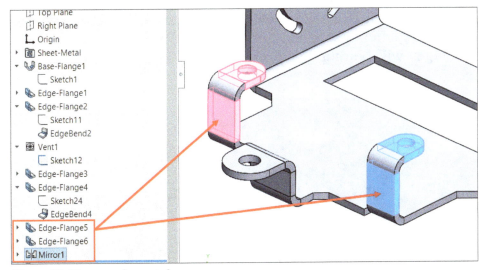

*Figure-53. Flanges to be created*

## Creating Flanges with Sketched Bend lines

- Click on the **Base Flange/Tab** tool from the **Sheet Metal CommandManager** in the **Ribbon**. You will be asked to select a planar face.
- Select flat face of the model and create sketch as shown in Figure-54.

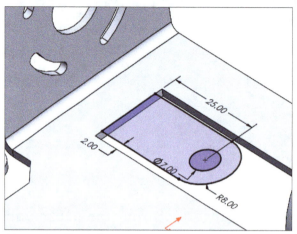

*Figure-54. Sketch for tab*

- After creating sketch, click on the **Exit Sketch** tool from the **Ribbon**. Preview of tab will be displayed; refer to Figure-55.

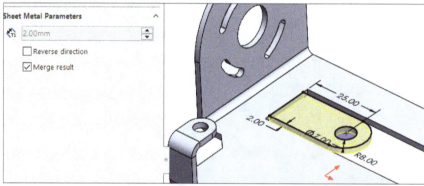

*Figure-55. Preview of tab feature*

- Click on the **OK** button from the **PropertyManager** to create the tab.
- Click on the **Sketched Bend** tool from the **Sheet Metal CommandManager** in the **Ribbon**. You will be asked to select planar face on which bend will be created.
- Select flat face of tab feature and create the sketch bend as shown in Figure-56. After creating line, click on the **Exit Sketch** tool from **Ribbon**. The **Sketched Bend PropertyManager** will be displayed and you will be asked to select fixed face for bending.

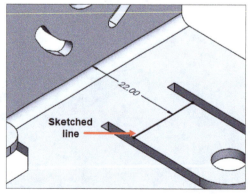

*Figure-56. Sketched line for bend*

- Select the flat face of model and set parameters as shown in Figure-57. Preview of bend will be displayed.

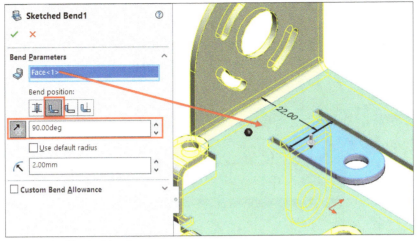

*Figure-57. Sketched bend parameters*

- Click on the **OK** button from **PropertyManager** to create the bend.
- Click again on **Sketched Bend** tool from the **Ribbon** and create sketched bend on rest portion of tab as shown in Figure-58.

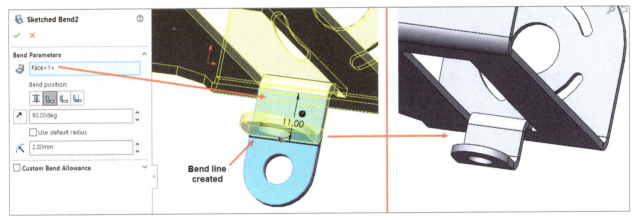

*Figure-58. Sketched bend created*

- Similarly, create tab and sketched bends on other side of model; refer to Figure-59.

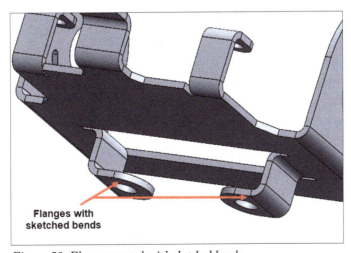

*Figure-59. Flanges created with sketched bends*

## Creating Bridge Lance

- If your software is not configured properly for forming tools then click on the **Options** button from **Quick Access Toolbar** at the top in application window. The **System Options** dialog box will be displayed.

- Select **File Locations** option from left box and select **Design Library** option from the **Show folders for** drop-down. The options will be displayed as shown in Figure-60.

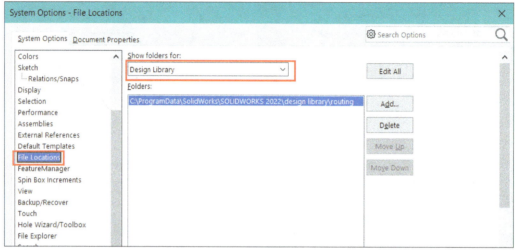

*Figure-60. System Options dialog box*

- Click on the **Add** button from the right area in dialog box. The **Select Folder** dialog box will be displayed.

- Select the folder **C:\ProgramData\SolidWorks\SolidWorks 20xx\design library\ form tools** from the dialog box and click on the **Select Folder** button. The folder will be added in the list; refer to Figure-61.

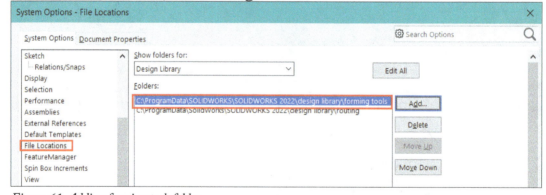

*Figure-61. Adding forming tools folder*

- After selecting folder, click on the **OK** button from the dialog box. A warning message box will be displayed. Select **Yes** button from the message box. The forming tools will be added in design library.

- Select the **Design Library** tab from task pane at the right in the application window. The **Design Library** task pane will be displayed; refer to Figure-62.

- Select the lances folder from forming tools category in the task pane and drag the bridge lance feature on the model from task pane; refer to Figure-63. On placing the feature, **Form Tool Feature PropertyManager** will be displayed; refer to Figure-64.

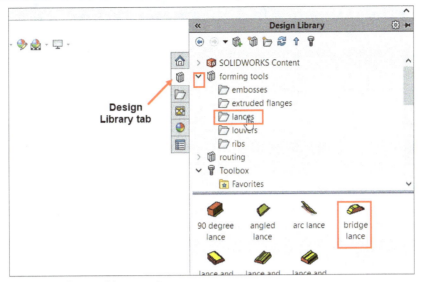

*Figure-62. Design library options*

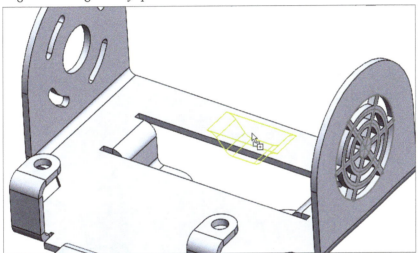

*Figure-63. Dragged bridge lance feature*

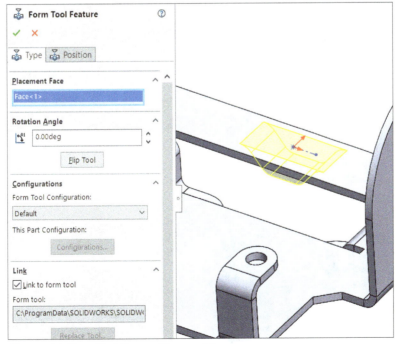

*Figure-64. Form Tool Feature PropertyManager*

- Click on the **Flip Tool** button from the **PropertyManager** to reverse direction of feature.
- Click on the **Position** tab from **PropertyManager**. The tools of **Sketch CommandManager** will become active.
- Using the **Smart Dimension** tool from the **Ribbon** and specify dimensions as shown in Figure-65.

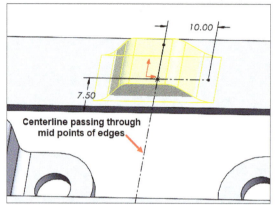

*Figure-65. Dimensioning insertion point*

- Click on the **OK** button from the **PropertyManager** to create the feature.
- Create a hole of diameter **4** at the center of flat face of lance feature using **Extruded Cut** tool or **Simple Hole** tool; refer to Figure-66.

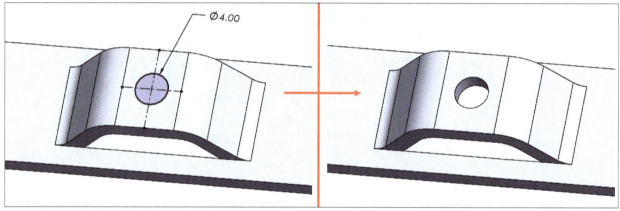

*Figure-66. Sketch for hole at flat face*

- Similarly, create other bridge lance feature with hole in the model; refer to Figure-67.

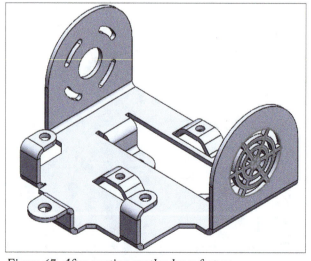

*Figure-67. After creating another lance feature*

# PRACTICAL 4 (ADVANCED LEVEL)

Create the sheetmetal model as shown in Figure-68.

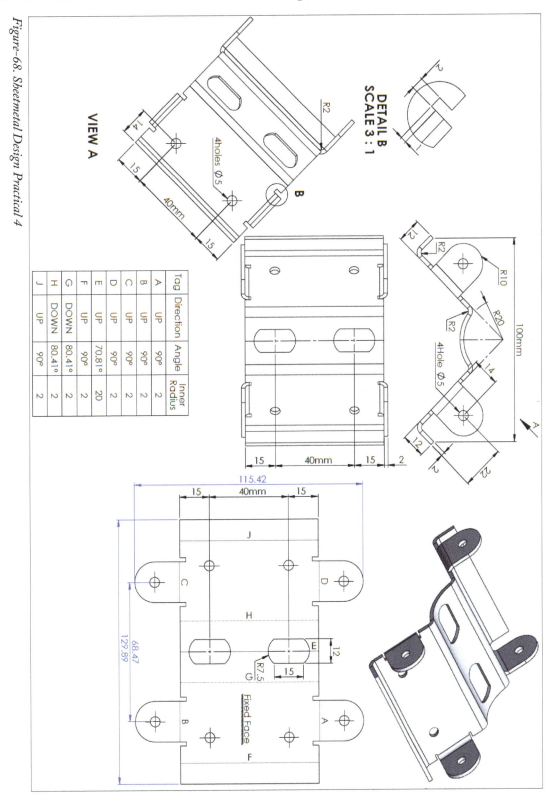

*Figure-68. Sheetmetal Design Practical 4*

| Tag | Direction | Angle | Inner Radius |
|-----|-----------|-------|--------------|
| A | UP | 90° | 2 |
| B | UP | 90° | 2 |
| C | UP | 90° | 2 |
| D | UP | 90° | 2 |
| E | UP | 70.81° | 20 |
| F | UP | 90° | 2 |
| G | DOWN | 80.41° | 2 |
| H | DOWN | 80.41° | 2 |
| J | UP | 90° | 2 |

## Observation on Drawing / Blueprint Reading

Thickness of the model is 2 mm and most of the features are sheetmetal features so clearly this part will be designed using Sheet Metal Design tools. Earlier, we have used **Base Flange/Tab** tool to create base tab but in this practical, we will use this tool to create a base flange. Rest of the features are edge flanges, extruded cuts, and holes.

## Starting a New Document

- Start SolidWorks if not started yet.
- Click on **New** tool from **File** menu or press **CTRL+N** using Keyboard. The **New SOLIDWORKS Document** dialog box will be displayed.
- Double-click on **Part** button from the dialog box. A new part document will open.
- Click on the **Sheet Metal CommandManager** from the **Ribbon** to display tools for sheet metal design.

## Creating Base Flange

- Click on the **Base Flange/Tab** tool from the **Sheet Metal CommandManager** in the **Ribbon**. You will be asked to select a plane for creating profile sketch.
- Select the **Front Plane** from graphics area. The sketching environment will become active.
- Create the sketch as shown in Figure-69 and click on the **Exit Sketch** button from the **Ribbon**. Preview of base flange will be displayed.

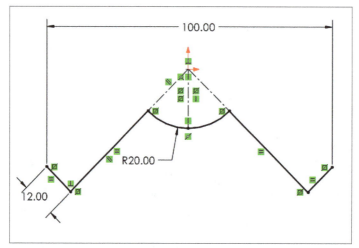

*Figure-69. Sketch for base flange*

- Set the parameters as shown in Figure-70 and click on the **OK** button from **PropertyManager**.

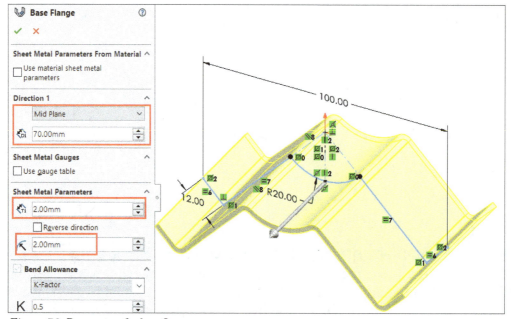

*Figure-70. Parameters for base flange*

## Creating Flange

- To create flanges in the model, you need to unfold all the bends because the dimensions for these flanges are given in flat pattern. Click on the **Unfold** tool from the **Sheet Metal CommandManager** in the **Ribbon**. The **Unfold PropertyManager** will be displayed.
- Set the parameters as shown in Figure-71 and click on the **OK** button to unfold the model.

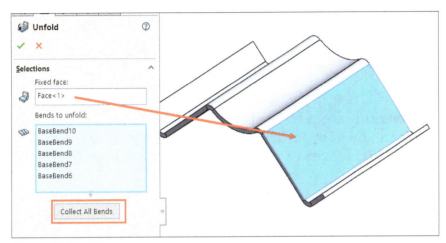

*Figure-71. Unfold parameters*

- Click on the **Edge Flange** tool from the **Ribbon**. The **Edge-Flange PropertyManager** will be displayed.
- Select the edge as shown in Figure-72 and set the parameters as per the figure in **PropertyManager**.

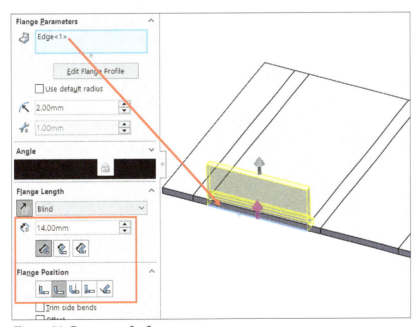

*Figure-72. Parameters for flange*

- Click on the **Edit Flange Profile** button from the **PropertyManager**. The sketching environment will become active.
- Create the sketch as shown in Figure-73 and then click on the **Finish** button.

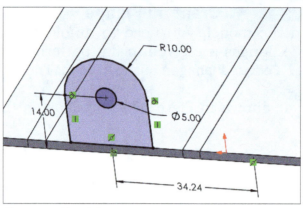

*Figure-73. Sketch for flanges*

## Creating Mirror Copies of Flanges

- Click on the **Plane** tool from the **Reference Geometry** drop-down in the **Features CommandManager** in the **Ribbon**. The **Plane PropertyManager** will be displayed.
- Select two parallel side faces of the model to create mid plane; refer to Figure-74.

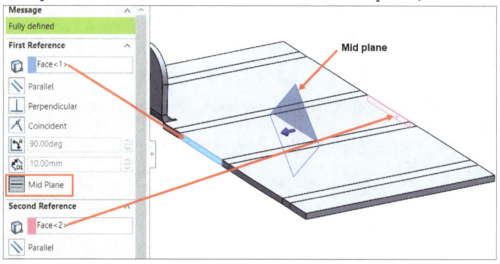

*Figure-74. Creating mid plane*

- Click on the **OK** button from the **PropertyManager** to create the plane.
- Similarly, create other mid plane; refer to Figure-75.

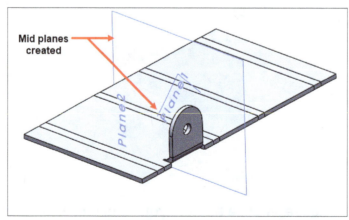

*Figure-75. Mid planes created*

- Select the **Edge-Flange1** feature from **FeatureManager Design Tree** created earlier and click on the **Mirror** tool from **Features CommandManager** in the **Ribbon**. The **Mirror PropertyManager** will be displayed.

- Select the **Plane1** and then **Plane2** from graphics area to be used as mirror planes. Preview of mirrored features will be displayed; refer to Figure-76.

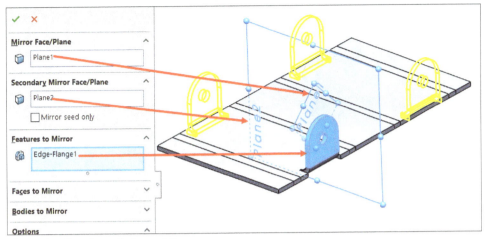

*Figure-76. Preview of mirrored features*

- Click on the **OK** button from the **PropertyManager** to create the mirror copies.

## Creating Extruded Cut Features

- Click on the **Extruded Cut** tool from the **Sheet Metal CommandManager** in the **Ribbon**. The **Extrude PropertyManager** will be displayed.
- Select top face of the model and create sketch as shown in Figure-77.

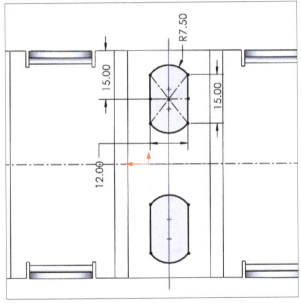

*Figure-77. Sketch for first extruded cut*

- Click on the **Exit Sketch** tool from **Ribbon**. The **Cut-Extrude PropertyManager** will be displayed.
- Set the parameters as shown in Figure-78 and click on the **OK** button. The extruded cut feature will be created.

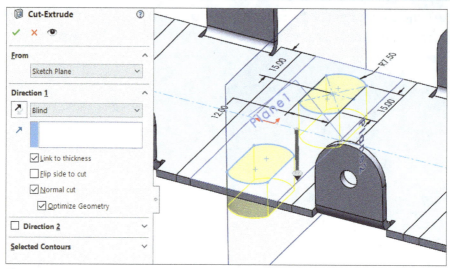

*Figure-78. Preview of Extrude cut*

- Click on the **Fold** tool from the **Sheet Metal CommandManager** in the **Ribbon**. The **Fold PropertyManager** will be displayed .
- Set the parameters as shown in Figure-79 and click on the **Collect All Bends** button.

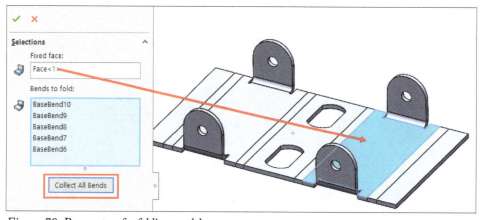

*Figure-79. Parameters for folding model*

## Creating Simple Holes

- Click on the **Simple Hole** tool from the **Sheet Metal CommandManager** in the **Ribbon**. The **Hole PropertyManager** will be displayed.
- Select flat face of the model earlier used as fixed face and place hole as shown in Figure-80.

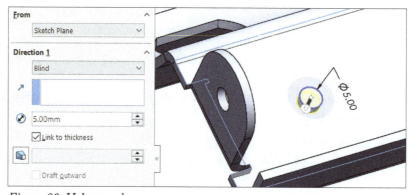

*Figure-80. Hole created*

- Click on the **OK** button from **PropertyManager** to create hole.

- Expand **Hole1** feature from **FeatureManager Design Tree** and right-click on sketch node from the **Design Tree**; refer to Figure-81.

*Figure-81. Right-click shortcut menu*

- Select the **Edit Sketch** button from the shortcut menu. The sketching environment will become active.
- Create the sketch as shown in Figure-82 and click on the **Exit Sketch** button. Two holes will be created.

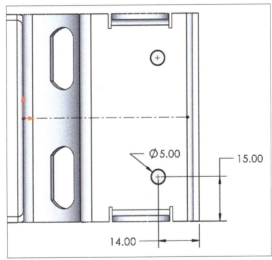

*Figure-82. Sketch for holes*

- Select **Hole1** from the **FeaturesManager Design Tree** and click on the **Mirror** tool from the **Features CommandManager** in the **Ribbon**. The **Mirror PropertyManager** will be displayed.
- Select **Right Plane** from the **FeaturesManager Design Tree** as mirror plane. Preview of mirrored holes will be displayed; refer to Figure-83.

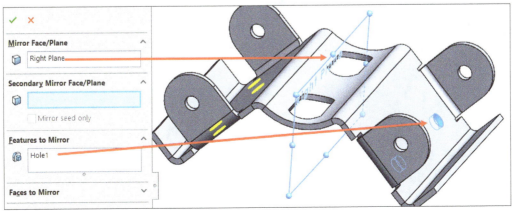

*Figure-83. Preview of mirrored holes*

- Click on the **OK** button from the **PropertyManager** to created mirrored copies. The final model will be displayed; refer to Figure-84.

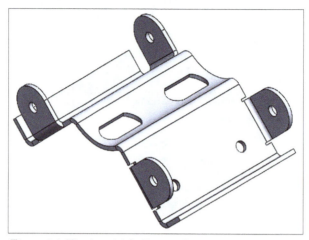

*Figure-84. Final model for Practical 4*

## PRACTICE 1 TO 6

Create the sheet metal drawings as shown in Figure-85 to Figure-90.

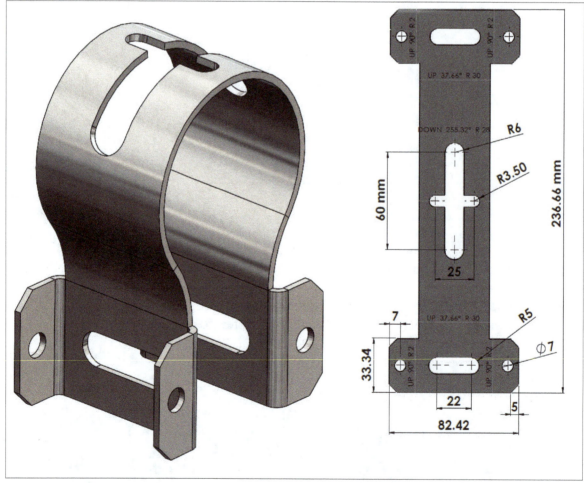

*Figure-85. Sheetmetal Design Practice 1*

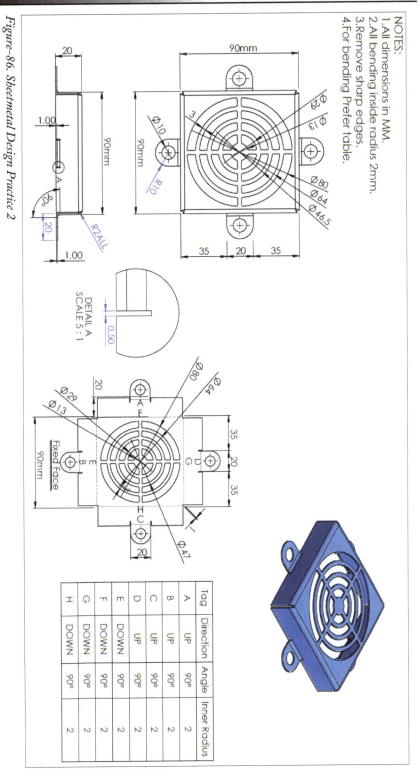

Figure–86. Sheetmetal Design Practice 2

NOTES:
1. All dimensions in MM.
2. All bending inside radius 2mm.
3. Remove sharp edges.
4. For bending Prefer table.

DETAIL A
SCALE 5 : 1

| Tag | Direction | Angle | Inner Radius |
|-----|-----------|-------|--------------|
| A | UP | 90° | 2 |
| B | UP | 90° | 2 |
| C | UP | 90° | 2 |
| D | UP | 90° | 2 |
| E | DOWN | 90° | 2 |
| F | DOWN | 90° | 2 |
| G | DOWN | 90° | 2 |
| H | DOWN | 90° | 2 |

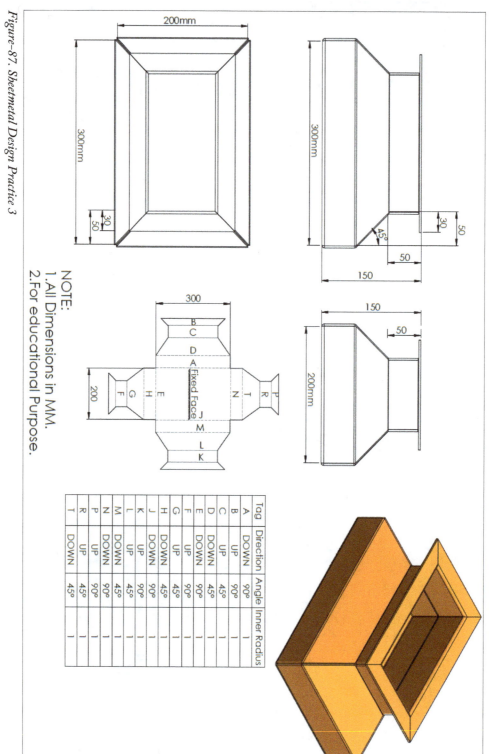

*Figure-87. Sheetmetal Design Practice 3*

NOTE:
1. All Dimensions in MM.
2. For educational Purpose.

| Tag | Direction | Angle | Inner Radius |
|-----|-----------|-------|--------------|
| A | DOWN | 90° | — |
| B | UP | 90° | — |
| C | UP | 45° | — |
| D | DOWN | 45° | — |
| E | DOWN | 90° | — |
| F | UP | 90° | — |
| G | UP | 45° | — |
| H | DOWN | 45° | — |
| J | DOWN | 90° | — |
| K | UP | 90° | — |
| L | UP | 45° | — |
| M | DOWN | 45° | — |
| N | DOWN | 90° | — |
| P | UP | 90° | — |
| R | UP | 45° | — |
| T | DOWN | 45° | — |

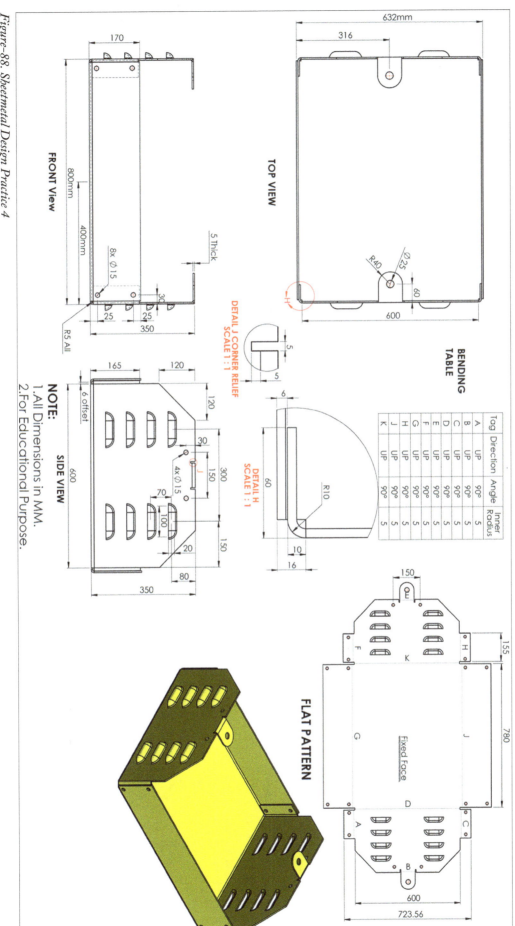

Figure-88. Sheetmetal Design Practice 4

**NOTE:**
1.All Dimensions in MM.
2.For Educational Purpose.

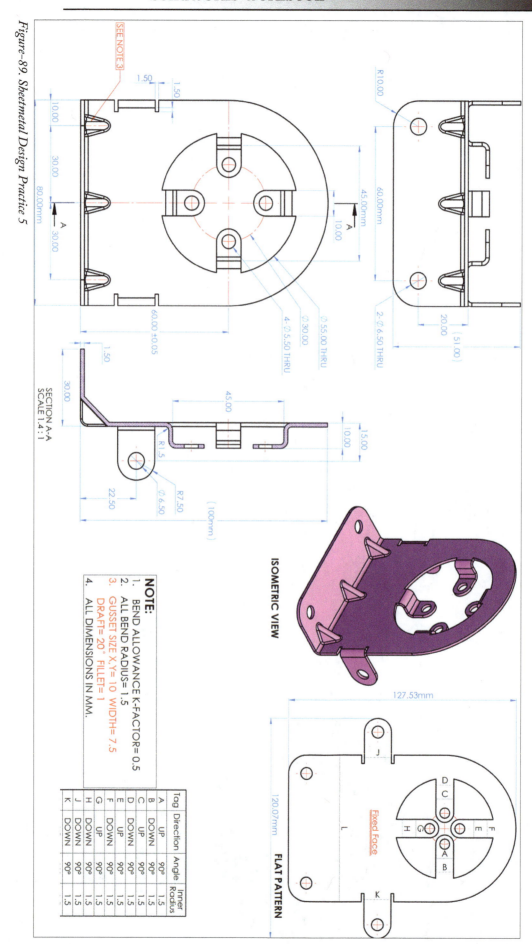

*Figure-89. Sheetmetal Design Practice 5*

ISOMETRIC VIEW

SECTION A-A
SCALE 1.4 : 1

FLAT PATTERN

**NOTE:**
1. BEND ALLOWANCE K-FACTOR= 0.5
2. ALL BEND RADIUS= 1.5
3. GUSSET SIZE X,Y= 10  WIDTH= 7.5
   DRAFT= 20°  FILLET= 1
4. ALL DIMENSIONS IN MM.

| Tag | Direction | Angle | Inner Radius |
|-----|-----------|-------|--------------|
| A | UP | 90° | 1.5 |
| B | DOWN | 90° | 1.5 |
| C | UP | 90° | 1.5 |
| D | DOWN | 90° | 1.5 |
| E | UP | 90° | 1.5 |
| F | DOWN | 90° | 1.5 |
| G | UP | 90° | 1.5 |
| H | DOWN | 90° | 1.5 |
| J | DOWN | 90° | 1.5 |
| K | DOWN | 90° | 1.5 |

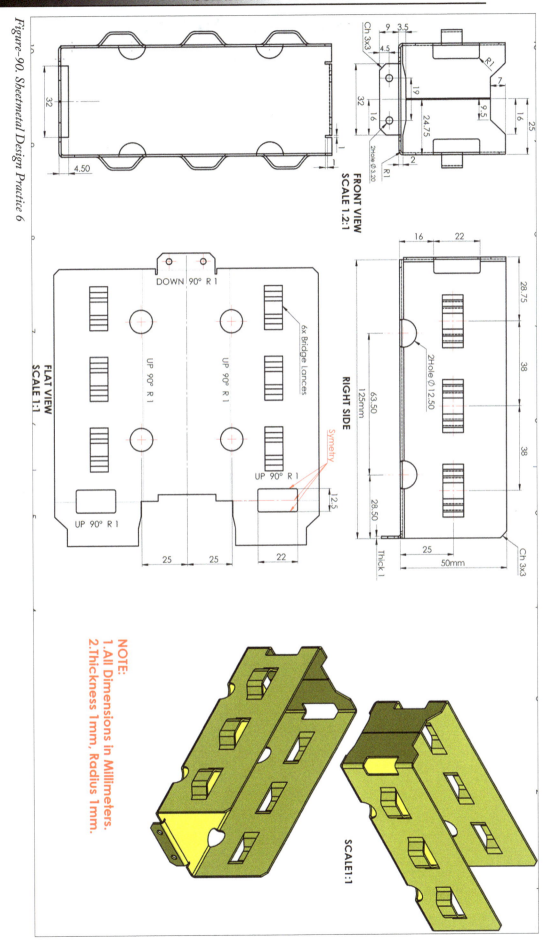

Figure-90. Sheetmetal Design Practice 6

**FRONT VIEW**
**SCALE 1.2:1**

**FLAT VIEW**
**SCALE 1:1**

**RIGHT SIDE**

DOWN 90° R 1

UP 90° R 1

UP 90° R 1

UP 90° R 1

UP 90° R 1

6x Bridge Lances

Symetry

2Hole Ø 12.50

Ch 3x3

Thick 1

**SCALE1:1**

**NOTE:**
**1.All Dimensions in Millimeters.**
**2.Thickness 1mm, Radius 1mm.**

# FOR STUDENT NOTES

FOR STUDENT NOTES

# Chapter 6

# Assembly Design
# Practical and Practice

## Topics Covered

The major topics covered in this chapter are:

- *Basic Assembly Design*
- *Advanced Assembly Design*

# INTRODUCTION

In this chapter, you will learn to create assembly design. An assembly is a combination of parts attached in defined ways to achieve a motion or function goal. For example, a ceiling fan is an assembly of multiple parts (components). There are two approaches to create assembly in most of the CAD software: Top-Down Approach and Bottom-Up Approach. Most commonly used approach is Bottom-Up approach in which all the components have already been created using Part Modeling environment and you will be placing these components according to their constraints in assembly file. The Top-Down approach is used when you are going to create all the components in assembly environment. In this method, you will place a blank component constrained to default references in assembly environment and then you will modify the component by accessing solid modeling tools. If you are developing a new assembly for production then a mixed approach is generally more suitable. In this approach, you will first create the components that are going to act as foundation for assembly and then you will create moving parts from assembly environment so that there is no issue of fitting of moving components. In this chapter, you will work on all the approaches to create assemblies.

# PRACTICAL 1 (BOTTOM UP APPROACH)

Create assembly as shown in Figure-1 using the components provided for this practical in resources of this book.

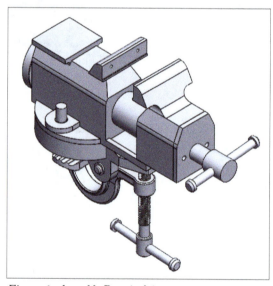

*Figure-1. Assembly Practical 1*

## Starting Assembly Design

- Start SolidWorks if not started yet.
- Click on the **New** tool from the **Quick Access Toolbar** or **File** menu. The **New SOLIDWORKS Document** dialog box will be displayed.
- Double-click on the **Assembly** button from dialog box. The assembly environment will become active and **Begin Assembly PropertyManager** will be displayed at the left in the application window; refer to Figure-2.

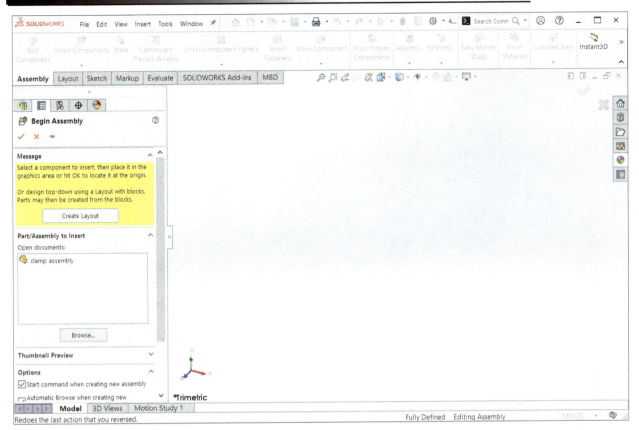

*Figure-2. Assembly environment*

- Click on the **Browse** button from the **PropertyManager**. The **Open** dialog box will be displayed.
- Select all the part files from the **Practical 1** resource folder while holding **CTRL** key; refer to Figure-3 and click on the **Open** button. All the selected parts will be displayed in the **Part/Assembly to Insert** section of **PropertyManager**.

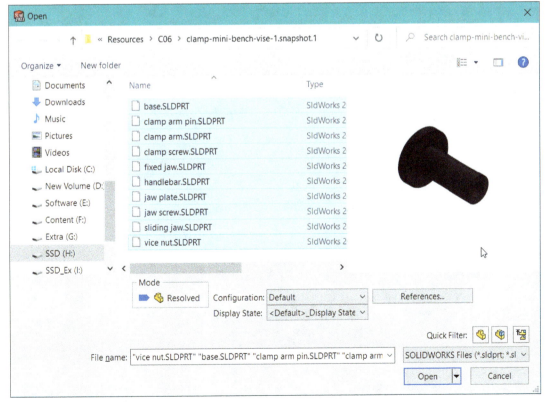

*Figure-3. Open dialog box*

- Note that by default the base part will get attached to cursor; refer to Figure-4.

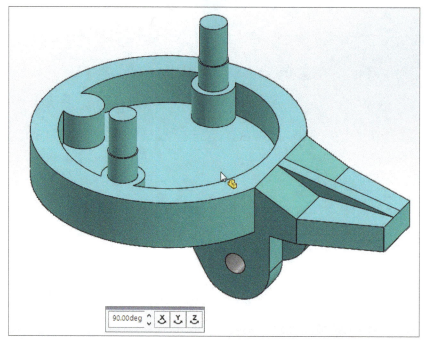

*Figure-4. Part attached to cursor*

- Click at desired location to place the base part which will act as base for other components and will be fixed. Note that **(f)** will be displayed with part name in **FeatureManager Design Tree** which designates that the part is fixed.
- On placing the base part, jaw part will get attached to cursor. Click in the graphics area to place the part and keep on placing all the other parts at adequate distance from each other; refer to Figure-5.

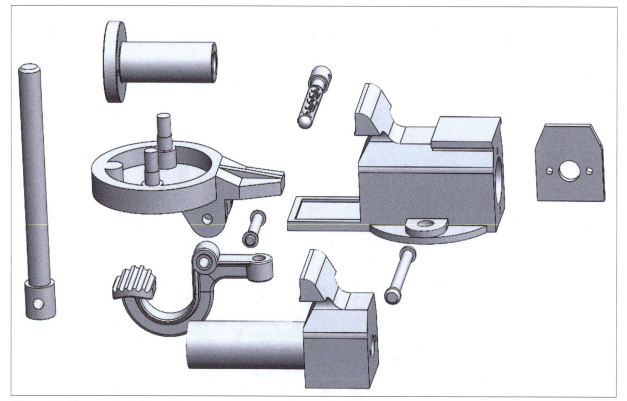

*Figure-5. All parts placed in graphics area*

## Applying Relationships (mates) to Components

- In this practical, we will hide all the components expect those on which we are working so that components are easily accessible. Select all the components except base and clamp arm from the **FeatureManager Design Tree** and right-click on any of the component. A shortcut menu will be displayed; refer to Figure-6.

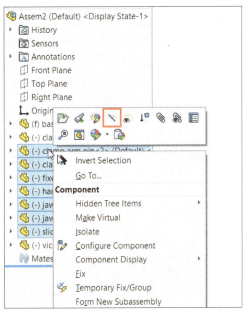

*Figure-6. Hide option*

- Select the **Hide Components** option from shortcut menu. Now only base and clamp arm will be displayed in graphics area; refer to Figure-7.

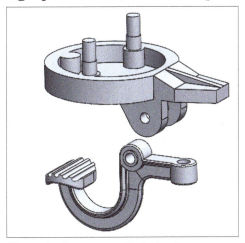

*Figure-7. Components displaying in graphics area*

- Select holes of the components as shown in Figure-8. A shortcut menu will be displayed.

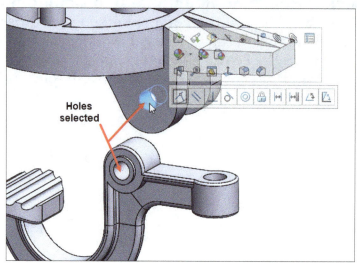

*Figure-8. Holes selected for mate*

- Select the **Concentric** button from toolbar in shortcut menu. The concentric mate will be applied to faces; refer to Figure-9.

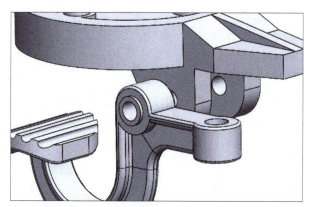

*Figure-9. After applying concentric mate*

- Click on the **Mate** tool from the **Assembly CommandManager** in the **Ribbon**. The **Mate PropertyManager** will be displayed; refer to Figure-10.

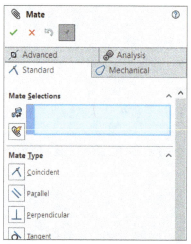

*Figure-10. Mate PropertyManager*

- Select the **Coincident** button from **Mate Type** rollout in **PropertyManager** and select the faces of model as shown in Figure-11. The selected faces will be attached to each other.

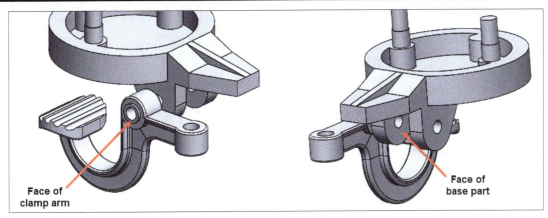

*Figure-11. Faces to select for coincident mate*

- Select the clamp arm pin component from **FeatureManager Design Tree** and click on the **Show Components** button from shortcut menu toolbar; refer to Figure-12. The component will display in graphics area.

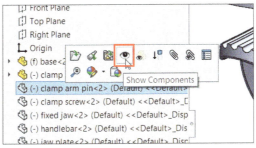

*Figure-12. Showing clamp arm pin*

- Click on the **Mate** tool from the **Ribbon**. The **Mate PropertyManager** will be displayed.
- Click on the **Coincident** button from **PropertyManager** and select the axis of hole and then axis of clamp arm pin; refer to Figure-13. A toolbar will be displayed on the model.
- Click on the **OK** button from the toolbar or **PropertyManager** to apply constraint.

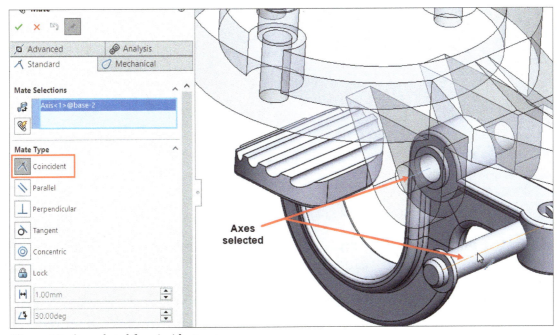

*Figure-13. Axes selected for coincident mate*

- Similarly, apply coincident constraint on flat faces of clamp arm pin and base part; refer to Figure-14.

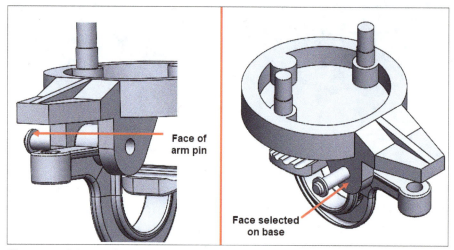

*Figure-14. Selecting faces for coincident constraint*

Note that you can check motion of various parts of assembly by selecting and dragging them in desired direction.

- In case of our model, if you move the clamp arm up and down by dragging then you will notice that the arm is passing through base part which is not desired. So, you need to apply angular limits. Click on the **Advanced** tab in the **PropertyManager**. The options in **PropertyManager** will be displayed as shown in Figure-15.

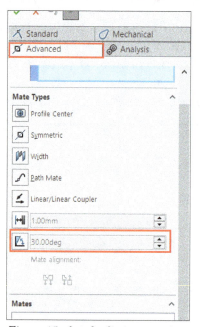

*Figure-15. Angular limit constraint*

- Click on the **Angle** button from the **PropertyManager**. The edit boxes for current angle position, maximum value, and minimum value will be displayed and you will be asked to select faces for applying constraint.
- Set the angle to 180 degree and select the faces as shown in Figure-16.

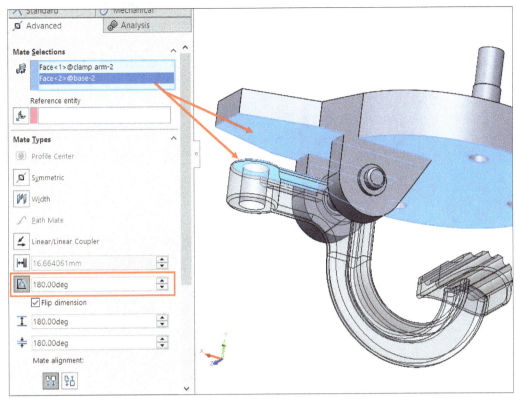

*Figure-16. Selecting faces for angle limit*

- Now, move the value in **Angle** edit box higher and lower to check the maximum and minimum limits of angular movement. We found the lower and upper angular limits 163 degree and 207 degree, respectively; refer to Figure-17.

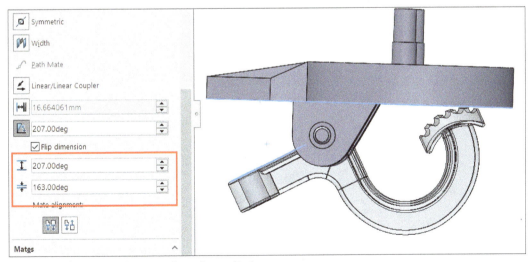

*Figure-17. Setting minimum and maximum limits*

- After setting the parameters, click on the **OK** button from **PropertyManager** to apply the limits. Now if you drag the clamp arm part in graphics area upward and downward, it will not pass through the base part.
- Click on the **Close** button from the **PropertyManager** to exit the **Mate PropertyManager**.
- Select the clamp screw from **Design Tree**, right-click on it, and select the **Show Components** button from shortcut menu toolbar. The part will become visible.

- Click on the **Mate** tool from **Ribbon** and select the **Coincident** button from **Mate Type** rollout of **PropertyManager**. You will be asked to select reference geometries.
- Select the axes of hole in clamp arm and shaft of clamp screw; refer to Figure-18. The two axes will get connected. Click on the **OK** button from **PropertyManager** to apply constraint.

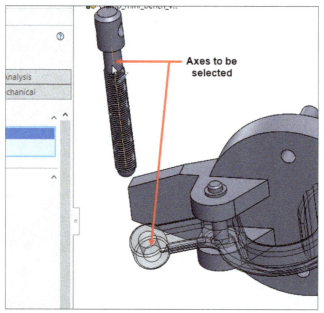

*Figure-18. Axes of screw and clamp arm*

- Click on the **Screw** button from **Mate Type** rollout in **Mechanical** tab of **PropertyManager** and select the faces as shown in Figure-19. The faces will get joined by screw constraint.

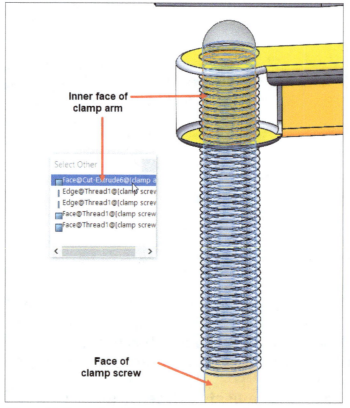

*Figure-19. Faces selected for screw mate*

- Set desired value for distance travelled in one revolution of screw; refer to Figure-20.

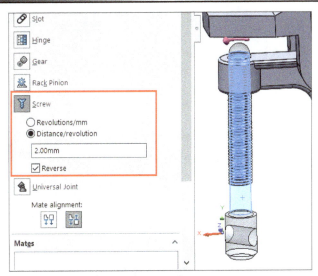

*Figure-20. Specifying parameters for screw*

- Click on the **OK** button from **PropertyManager** to apply the constraint. At the moment, we will leave this part free to move wildly. Later in this practical, we will create a part to restrict unwanted motion of screw. Click on the **Close** button from **PropertyManager** to exit.
- Select the fixed jaw part from **FeatureManager Design Tree** and set it to display.
- Click on the **Mate** tool from **PropertyManager** and apply constraints as shown in Figure-21.

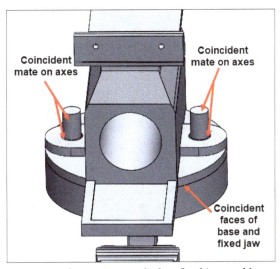

*Figure-21. Constraints applied on fixed jaw and base*

- Make the handlebar part display in graphics area and apply coincident axes constraint; refer to Figure-22.

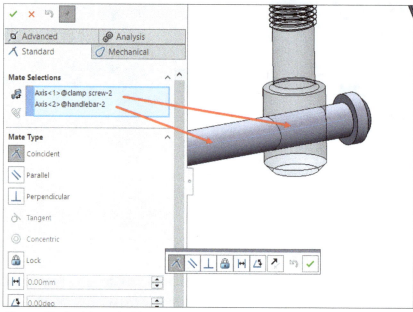

*Figure-22. Applying coincident axes constraint*

- Expand the Clamp screw and Handlebar parts in **In-graphics Design Tree** and apply the Advanced Distance constraint as shown in Figure-23. This constraint will keep handle bar attached to screw.

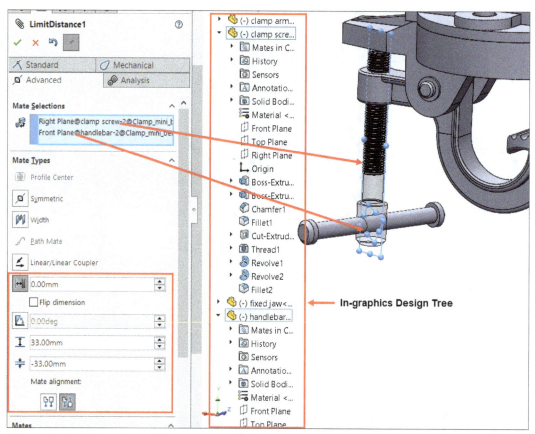

*Figure-23. Applying advanced distance constraint*

- Make the sliding jaw part visible from **In-graphics Design Tree**; refer to Figure-24.

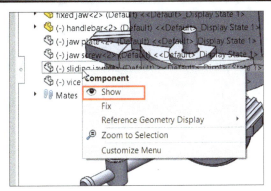

*Figure-24. Show option in right-click shortcut menu*

- Make the shaft of sliding jaw and hole of fixed jaw concentric; refer to Figure-25.

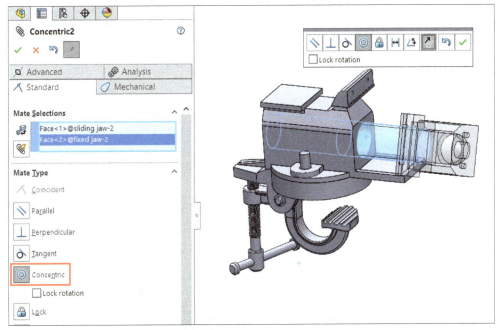

*Figure-25. Making sliding jaw and fixed jaw concentric*

- Make the faces of sliding and fixed jaws parallel as shown in Figure-26.

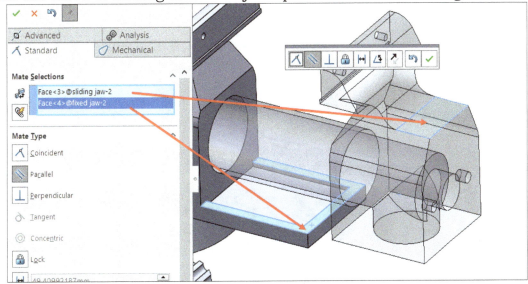

*Figure-26. Making faces parallel*

- Apply the advanced distance constraint to define distance limits for sliding jaw so that it do not pass through fixed jaw; refer to Figure-27. Click on the **OK** button from **PropertyManager** to apply mate.

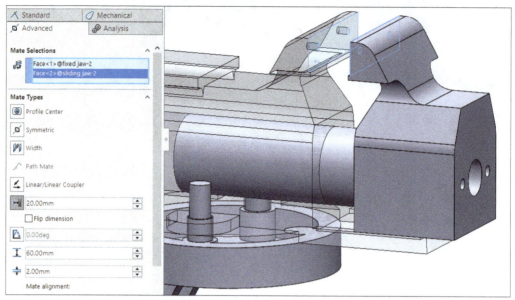

*Figure-27. Advanced distance constraint parameters*

- Right-click on jaw plate from **In-graphics Design Tree** and set it to show; refer to Figure-28.

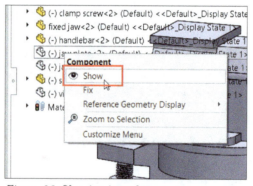

*Figure-28. Showing jaw plate*

- Make the holes of jaw concentric with sliding jaw holes; refer to Figure-29.

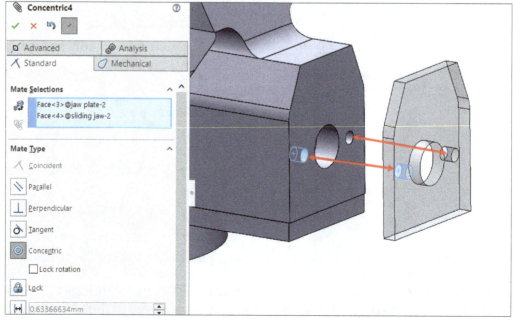

*Figure-29. Applying concentric constraints*

- Make the faces of jaw plate and sliding jaw coincident as shown in Figure-30.

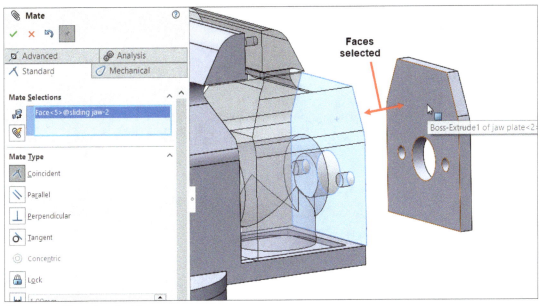

*Figure-30. Faces for coincident*

- Make the jaw screw visible using **In-graphics Design Tree** as discussed earlier.
- Make the axes of jaw screw and sliding jaw coincident; refer to Figure-31.

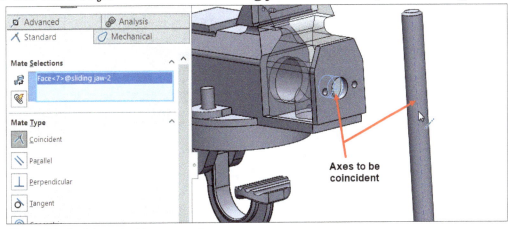

*Figure-31. Applying coincident constraint*

- Similarly, apply coincident constraint to flat faces of jaw screw and jaw plate; refer to Figure-32. The model will be displayed as shown in Figure-33. Close the **Mate PropertyManager**.

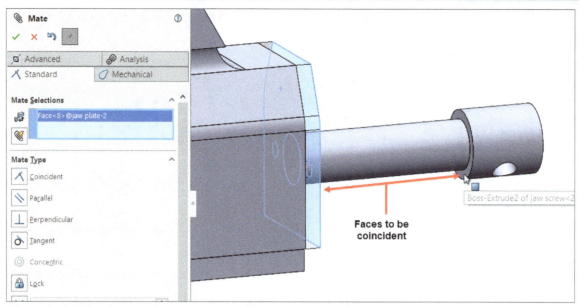

*Figure-32. Making faces coincident*

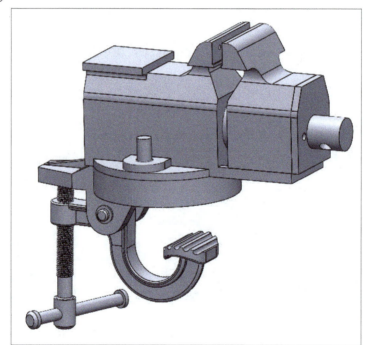

*Figure-33. Model after constraining screw*

- Now, we need another copy of handlebar. To do so, hold the **CTRL** key and drag the handlebar from **FeaturesManager Design Tree** in graphics area. The handlebar will be attached to cursor; refer to Figure-34.
- While holding the **CTRL** key, release the handle bar in graphics area and then release the **CTRL** key. The handlebar will be placed in graphics area.

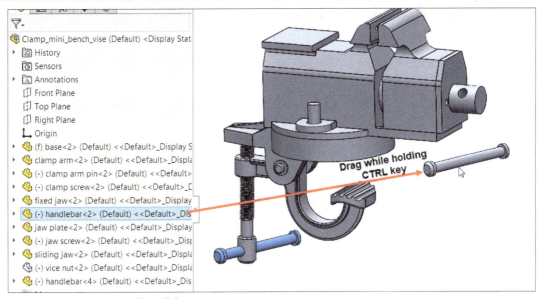

*Figure-34. Placing copy of handlebar*

- Apply the same constraints between handlebar and jaw screw as discussed for handlebar and clamp screw; refer to Figure-35.

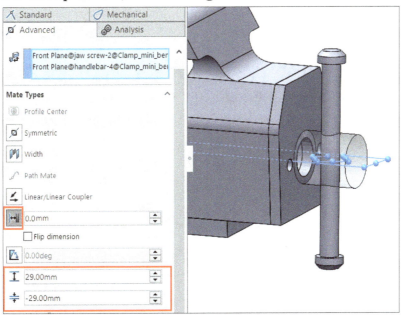

*Figure-35. Applying distance limits to handlebar*

- Set the last component which is vice nut to display as discussed earlier and make the holes concentric with holes on back side of fixed jaw; refer to Figure-36.

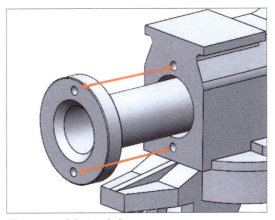

*Figure-36. Mating holes*

- Make the flat faces of fixed jaw and vice nut coincident. The final model will be displayed; as shown in Figure-37.

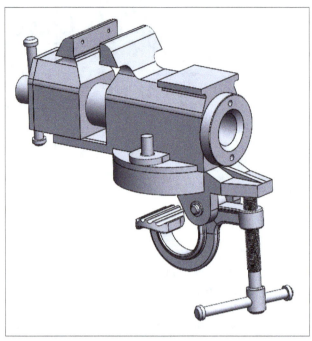

*Figure-37. Final assembly model*

## PRACTICAL 2 (TOP-DOWN APPROACH)

Create the assembly as shown in Figure-38. You will get a base part only as shown in Figure-39. You need to create fitting parts within the assembly.

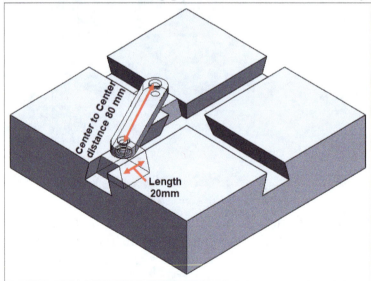

*Figure-38. Final assembly for Practical 2*

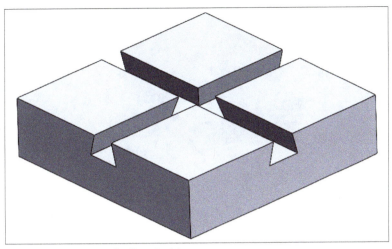

*Figure-39. Base part for Practical 2*

## Creating New Assembly

- Start SolidWorks if not started yet.
- Click on the **New** tool from the **Quick Access Toolbar** or **File** menu. The **New SOLIDWORKS Document** dialog box will be displayed.
- Double-click on the **Assembly** button from dialog box. The assembly environment will become active and **Begin Assembly PropertyManager** will be displayed at the left in the application window. Also, the **Open** dialog box will be displayed.
- Double-click on the Base part file from the Practical 2 folder of C06 directory in resource files. The part will get attached to cursor; refer to Figure-40.

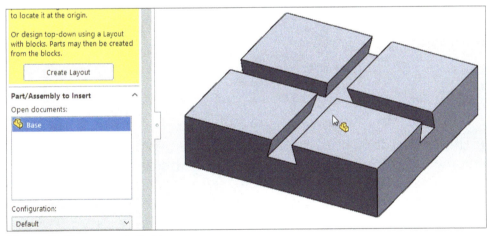

*Figure-40. Base part attached to cursor*

- Click at desired location to place the model. Note that assembly planes are not aligned with planes of base part; refer to Figure-41. To make these planes coincident, we need to first make the part float which is fixed by default.

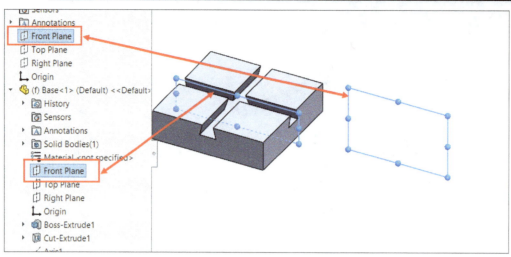

*Figure-41. Front Planes of base part and assembly*

- Right-click on **Base** part from the **Design Tree** and select the **Float** option from the shortcut menu. Now, the base part will be free to move.
- Select Front Planes of assembly and Base part from **Design Tree** while holding **CTRL** key and then release the **CTRL** key without moving cursor. The mate toolbar will be displayed; refer to Figure-42.

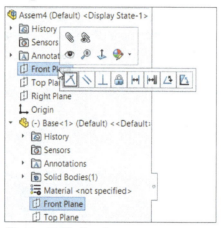

*Figure-42. Toolbar for mate*

- Click on the **Coincident** button to make the planes coincident. Similarly, make the Top Planes and Right Planes of base part and assembly; refer to Figure-43. The benefit of making planes coincident is easy reference selection for creating parts in assembly environment which you will learn later in this practical.

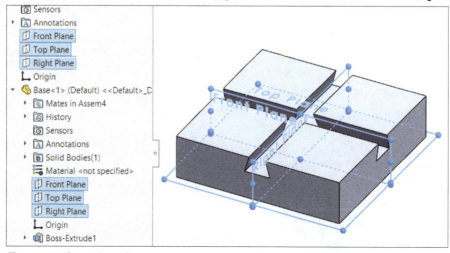

*Figure-43. Coincident planes*

## Creating Slider Part

- Click on the **New Part** tool from the **Insert Components** drop-down in the **Assembly CommandManager** of the **Ribbon**. You will be asked to select a plane/face to be used as placement reference for the part.

- Select the **Front Plane** of assembly environment. The **Edit Component** environment will become active with sketching mode; refer to Figure-44.

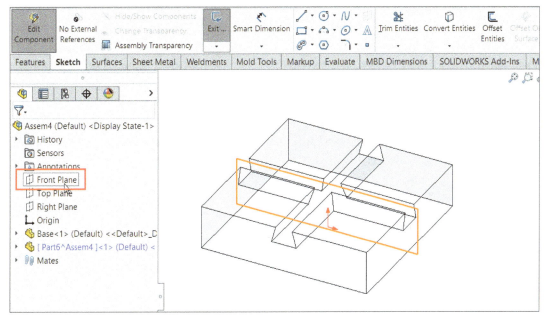

*Figure-44. Edit Component environment*

- Make the sketching plane parallel to screen by pressing **CTRL+8** (not from Numpad in keyboard).

- Click on the **Convert Entities** tool from the **Ribbon** and select the edges as shown in Figure-45. Click on the **OK** button from **PropertyManager**. The edges will be converted to lines in sketch.

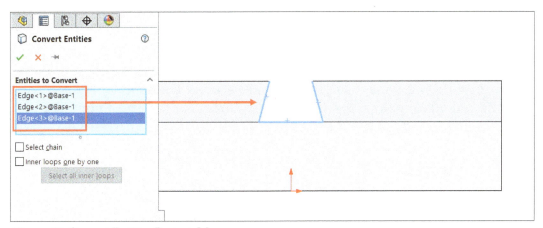

*Figure-45. Convert Entities PropertyManager*

- Click on the **Line** tool from the **Ribbon** and create a line joining end points of converted edges; refer to Figure-46.

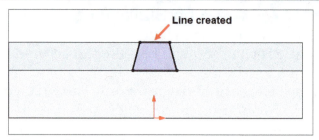

*Figure-46. Line created*

- Click on the **Exit Sketch** tool from graphics area or **Ribbon**. Click on the **Features CommandManager** in the **Ribbon** and select the **Extruded Boss/Base** tool from the **Ribbon**. The **Extrude PropertyManager** will be displayed.
- Select any line of the sketch recently created. Preview of extrude feature will be displayed.
- Select the **Mid Plane** option from **End Condition** drop-down in **Direction 1** rollout and set the **Depth** as **30** mm; refer to Figure-47.

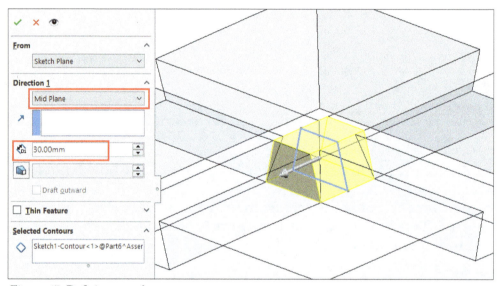

*Figure-47. Defining extrude parameters*

- Click on the **OK** button from **PropertyManager** to create the extrude feature.
- Click on the **Revolved Boss/Base** tool from **Features CommandManager** in the **Ribbon**. You will be asked to select a sketch plane/face. Select the **Front Plane** of part being created from **In-graphics Design Tree**; refer to Figure-48.

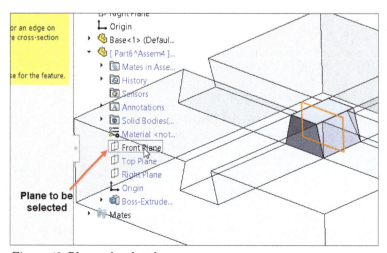

*Figure-48. Plane to be selected*

- Create the sketch as shown in Figure-49 to create boss feature for link attachment and click on the **Exit Sketch** tool from **Ribbon**. Preview of revolve feature will be displayed; refer to Figure-50.

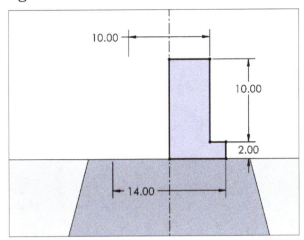

*Figure-49. Sketch for revolve*

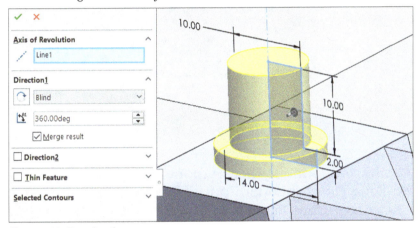

*Figure-50. Revolve feature*

- Click on the **OK** button from the **PropertyManager** to create the feature. Click on the **Edit Component** tool from **Ribbon** to exit part editing environment and go back to assembly environment.
- If you now try to drag the newly created component from its place then you will find that it is fixed which is not desirable for assembly. To free this component from its place, select the **InPlace1** constraint from **Mates** node in **Design Tree**; refer to Figure-51 and delete it by pressing **DELETE** key from keyboard. Note that on deleting the constraint, a warning message is displayed as shown in Figure-52.

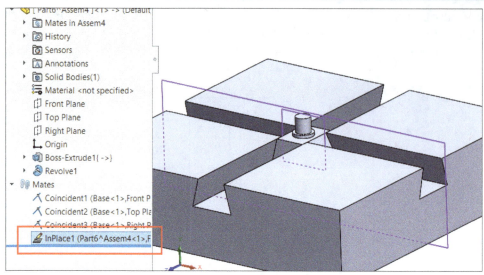

*Figure-51. InPlace constraint*

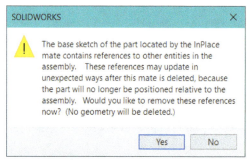

*Figure-52. Warning message*

- Click on the **Yes** button to make the component free to move so that we can apply desired mates. Drag the component out of base component.
- Click on the **Mate** tool from **Assembly CommandManager** in the **Ribbon**. The **Mate PropertyManager** will be displayed.
- Select the faces as shown in Figure-53 to apply mate. The coincident mate will get selected automatically in **PropertyManager** and toolbar. Click on the **OK** button to apply mate.
- Similarly, apply coincident constraint on bottom face of part and flat face in groove on base; refer to Figure-54.

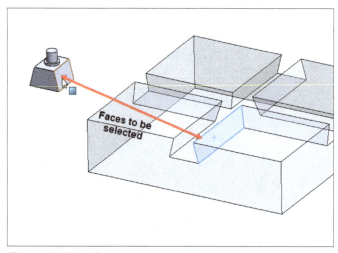

*Figure-53. Faces for mate*

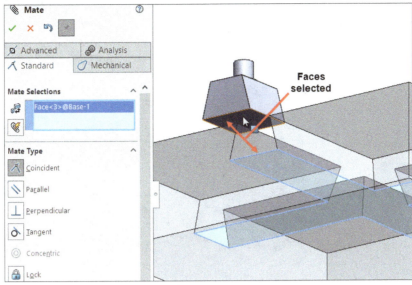

*Figure-54. Faces selected for coincident mate*

- Click on the **Close** button to exit the **Mate PropertyManager**.

## Copying Slider Part with Constraints

- Right-click on the new part from **FeatureManager Design Tree** and select the **Rename tree item** option to change name of part. The name of part will become available to edit. Type the name as **Slider** and press **ENTER**. The name of part will get changed.
- Right-click on this part again from **Design Tree** and select the **Copy with Mates** option to create copy of part; refer to Figure-55. The **Copy with Mates PropertyManager** will be displayed; refer to Figure-56.

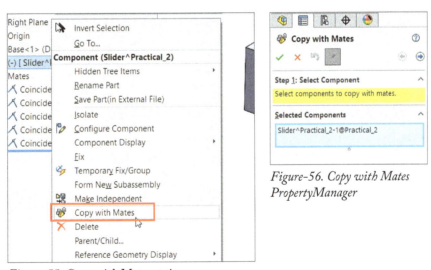

*Figure-56. Copy with Mates PropertyManager*

*Figure-55. Copy with Mates option*

- Click on the **Next** button from the **PropertyManager**. The options to define coincident constraints will be displayed; refer to Figure-57.

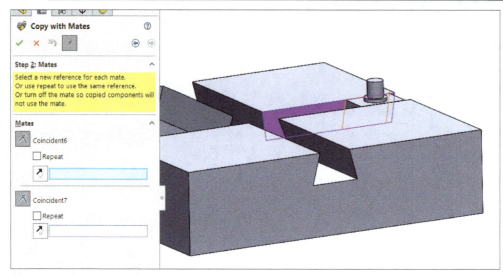

*Figure-57. Options to define coincident constraint*

- Select the faces as shown in Figure-58 to apply respective mates. The preview of placed component will be displayed.

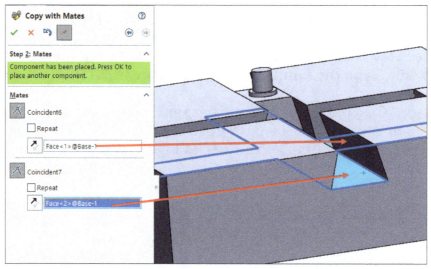

*Figure-58. Faces selected for mates*

- Click on the **OK** button from the **PropertyManager** to create the mates. Click on the **Close** button to exit the **PropertyManager**. The model will be displayed as shown in Figure-59.

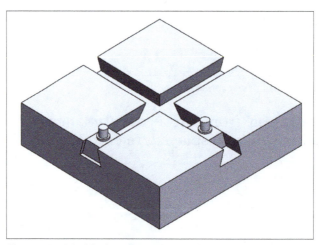

*Figure-59. Model after placing both sliders*

## Creating Link Part

- Click on the **New Part** tool from the **Insert Components** drop-down in the **Ribbon**. You will be asked to select a reference face for creating the component.
- Select flat face of slider part as shown in Figure-60. The sketching environment for new part will be activated.

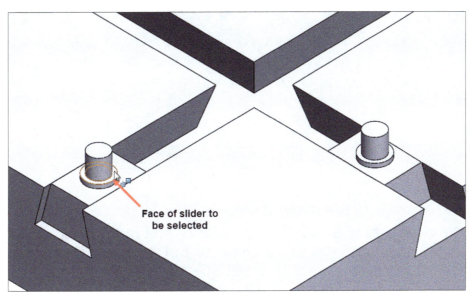

*Figure-60. Face selected for creating link*

- Click on the **Convert Entities** tool from the **Ribbon** and project the circles of slider parts as shown in Figure-61. Click on the **OK** button from **PropertyManager** to create projected sketch entities.

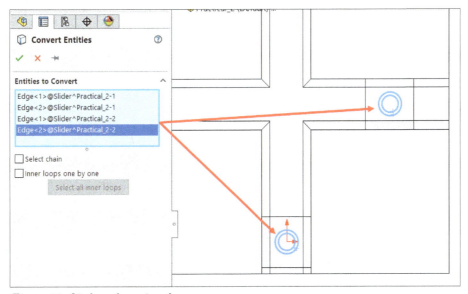

*Figure-61. Circles to be projected*

- Click on the **Line** tool and create two lines joining outer circles as shown in Figure-62. Make sure the lines are tangent to circles.

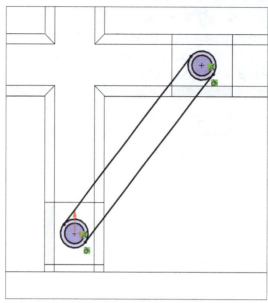

*Figure-62. Lines to be created*

- Click on the **Trim Entities** tool from the **Ribbon** and remove extra portion of outer circles as shown in Figure-63. Click on the **OK** button to exit the **PropertyManager**.

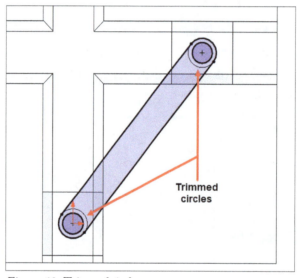

*Figure-63. Trimmed circles*

- Click on the **Exit Sketch** tool from the **Ribbon** to exit sketching environment.
- Click on the **Extruded Boss/Base** tool from **Features CommandManager** in the **Ribbon**. The **Extrude PropertyManager** will be displayed.
- Select any of the line of sketch. The parameters in **Extrude PropertyManager** will become active.
- Clear selected sections from **Selected Contours** rollout and then click in the middle of sketch to select section as shown in Figure-64.

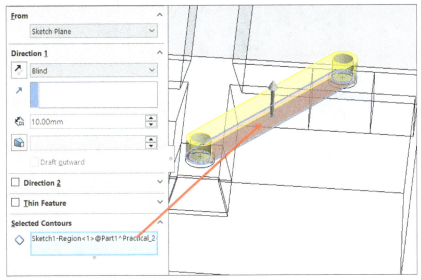

*Figure-64. Sketch section selected for extrude*

- Set the height of extrude feature to 8 mm and click on the **OK** button from the **PropertyManager** to create extrude feature.
- Click on the **Edit Component** tool to exit the editing mode.
- Delete the **InPlace** mate of newly created part as discussed earlier from the **Mates** node in **Design Tree**. The link part will become free to move.
- Apply the concentric and coincident constraints between sliders and link to place the part properly; refer to Figure-65.

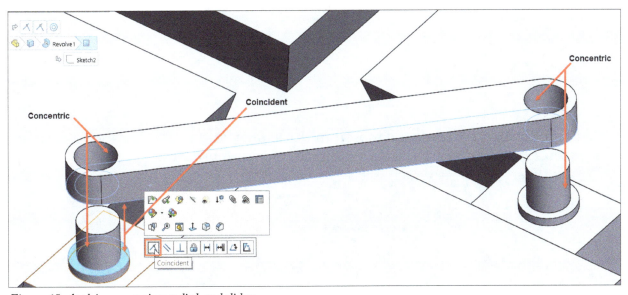

*Figure-65. Applying constraints to link and sliders*

- Now, only one parameter is left to define which is length of link part. Here, you will get to see the bidirectional associativity of the software in different environments. Right-click on Part1 which is link part from **Design Tree** and click on the **Open** option from shortcut menu toolbar; refer to Figure-66. The model will be displayed in Part environment.

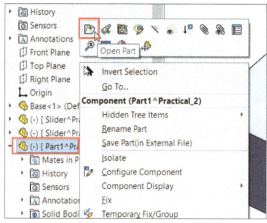

*Figure-66. Open Part option*

- Expand the Boss-Extrude1 feature of part from **Design Tree**, right-click on Sketch1 feature, and select the **Edit Sketch** option from shortcut menu toolbar; refer to Figure-67. The sketching environment will become active.

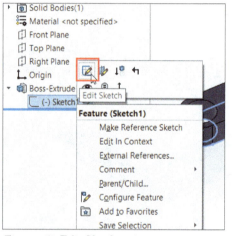

*Figure-67. Edit Sketch option*

- Click on the **Smart Dimension** tool from the **Ribbon** and apply distance dimension between centers of two circles as 80 mm; refer to Figure-68.

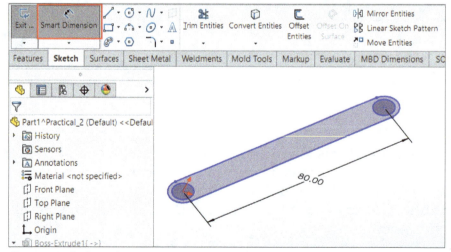

*Figure-68. Applying dimension*

- Click on the **Exit Sketch** tool from the **Ribbon**. The solid model of part will be displayed.

- Save the model by pressing **CTRL**+**S** and close the model document. A message box will be displayed defining that assembly has been changed based on modifications of part model; refer to Figure-69.

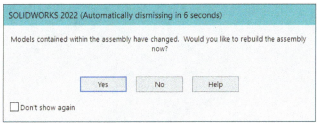

*Figure-69. Message box*

- Click on the **Yes** button to rebuild the assembly model to show updated parts. Press **CTRL**+**S** to save the model. The **Save As** dialog box will be displayed; refer to Figure-70.

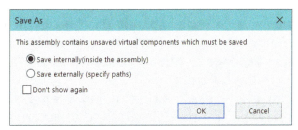

*Figure-70. Save As dialog box*

- Select the **Save internally** radio button to save the part file of components within the assembly file. This will increase the size of assembly file and make the parts not usable outside the assembly.
- Select the **Save externally** radio button to save the part files at desired locations in hard drive. This is suitable option for reusing the part files for other assemblies. Select the **Save externally** radio button and click on the **OK** button. The part files will be saved in same location where you have earlier saved the assembly file.

## Creating Exploded View

- Exploded view is used to check how the parts are assembled together. To create exploded view of assembly, click on the **Exploded View** tool from the **Assembly CommandManager** in the **Ribbon**. The **Explode PropertyManager** will be displayed.
- Select the components that are to be moved from the graphics area, click in the **Explode Direction** selection box of **PropertyManager**, and then select top flat face of base component to define direction of explode; refer to Figure-71.

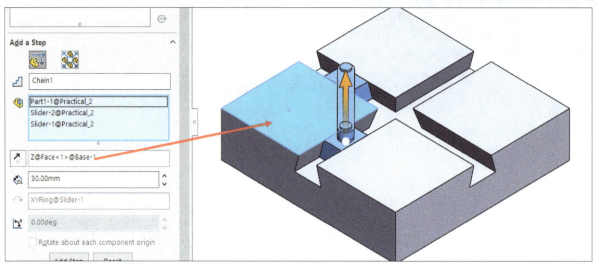

*Figure-71. Defining explode direction*

- Specify desired distance value in the **Explode Distance** edit box of **PropertyManager** to define gap between various components of assembly on explosion.
- Select the **Auto-space components** check box to make all the components distant by specified values from each other.
- After setting desired parameters, click on the **Add Step** button from the **Add a Step** rollout in **PropertyManager**. The model will be displayed as shown in Figure-72.

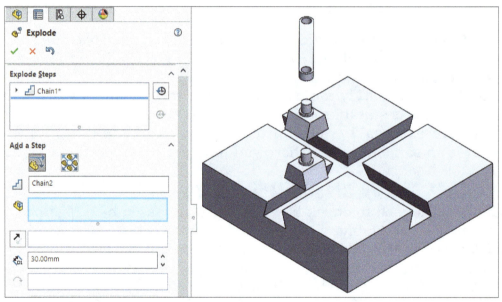

*Figure-72. After applying explode step*

- Click on the **OK** button from the **PropertyManager** to create exploded view configuration.
- To move back to collapsed view of assembly, right-click on the **Exploded View1** feature from **Default** node in the **ConfigurationManager**. A shortcut menu will be displayed; refer to Figure-73.

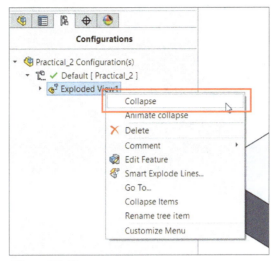

*Figure-73. Shortcut menu*

- Click on the **Collapse** option from the shortcut menu to return to collapsed view from exploded view of assembly. You can display exploded view again by using the same shortcut menu.

# PRACTICE 1

Create the assembly of differential using the part files provided in resources folder. The exploded view of model is shown in Figure-74. The assembled model will be displayed as shown in Figure-75.

*Figure-74. Exploded view Practice 1*

*Figure-75. Assembled Model Practice 1*

# PRACTICE 2

Create the chain sprocket assembly as shown in Figure-76. Make sure the chain and gears rotate in relation to each other. The components provided for this assembly are as shown in Figure-77.

*Figure-76. Assembled Model Practice 2*

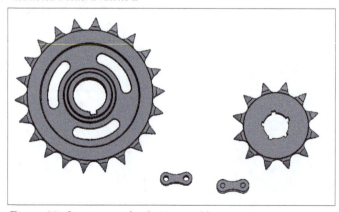

*Figure-77. Components for chain assembly*

(Hint: You will need **Chain Component Pattern** tool and **Belt/Chain** tool for this assembly.)

## PRACTICE 3

Create the shocker spring assembly as shown in Figure-78. The exploded view of assembly is shown in Figure-79.

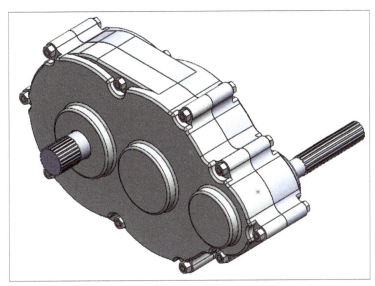

| ITEM NO. | PART NUMBER | DESCRIPTION | QTY. |
|----------|-------------|-------------|------|
| 1 | Spring | | 1 |
| 2 | Base Nut | | 1 |
| 3 | Spring Rod | | 1 |
| 4 | Base | | 1 |

*Figure-78. Assembled Model Practice 3*

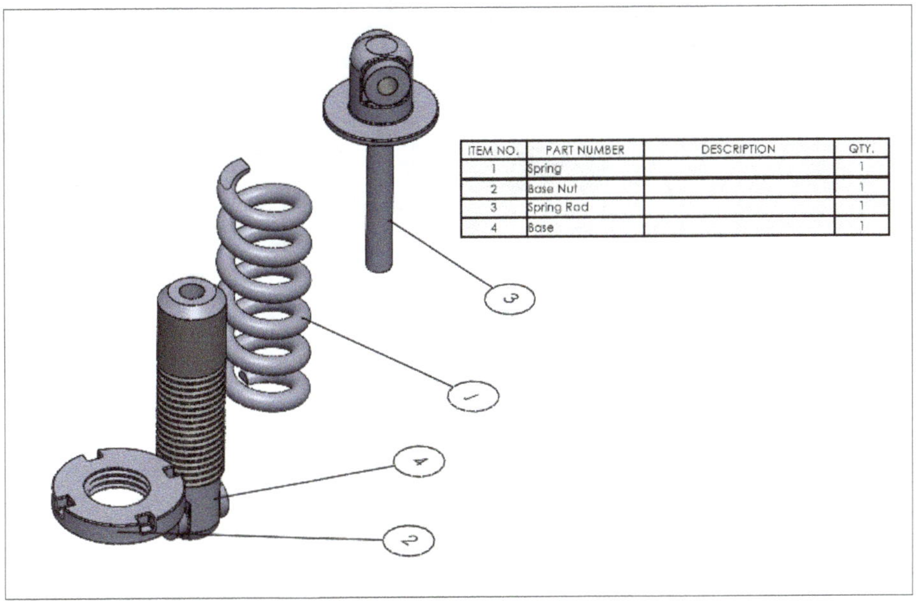

*Figure-79. Exploded view Practice 3*

## PRACTICE 4

Create gear box assembly as shown in Figure-80. The open view of assembly is shown in Figure-81. Exploded view of assembly is shown in Figure-82.

*Figure-80. Assembled Model Practice 4*

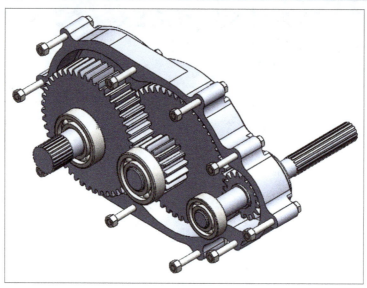

*Figure-81. Gear box assembly open*

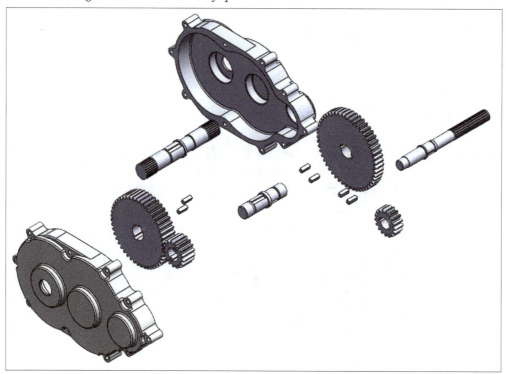

*Figure-82. Exploded view Practice 4*

(Hint: You will need SolidWorks Toolbox for inserting bearings, fasteners, and retaining rings.)

# FOR STUDENT NOTES

# FOR STUDENT NOTES

# Chapter 7

# Drafting Practical and Practice

## Topics Covered

The major topics covered in this chapter are:

- *Basic Drafting*
- *Advanced Drafting*

# INTRODUCTION

Drafting is the stage in product design at which you create 2D representation of 3D models so that they can be manufactured/assembled in the workshop. At this stage, you apply dimensions, annotations, and information related to manufacturing like material, finish, manufacturing processes used, and so on. Although, Model Based Annotations are available to annotate 3D model in portable document formats that can be view in various electronic gadgets but there is no replacement of good old method of printing drawings on paper for workshop. Note that there is just one rule for creating engineering drawings: there must be no uncertainty left in drawing for manufacturing a product. It means each feature must be fully defined by dimensions and tolerances. Various practical and practices on creating drawings are given next.

# PRACTICAL 1 (BASIC LEVEL)

Create drawing for the model shown in Figure-1.

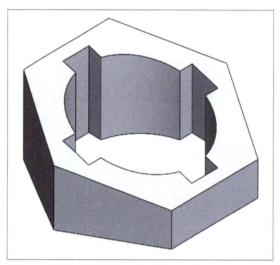

*Figure-1. Model for Drawing Practical 1*

Steps:
- Open the model in SolidWorks and make sure all the dimensions have been applied to the model with desired tolerances. This step ensures that the model is fully defined and you can directly import dimensions of model in drawing.
- Start a new drawing from Part Modeling environment and place the number of views that can fully define the model.
- Import the dimensions and apply more annotations if required.
- Define Title block parameters as desired and Save the file.

## Verifying Model Dimensions and Starting New Drawing Document
- Start SolidWorks if not started yet.
- Click on the **Open** tool from **Quick Access Toolbar** or press **CTRL+O**. The **Open** dialog box will be displayed.
- Select the **Practical 1** model file from resources folder of this chapter. The model will be displayed as shown in Figure-2.

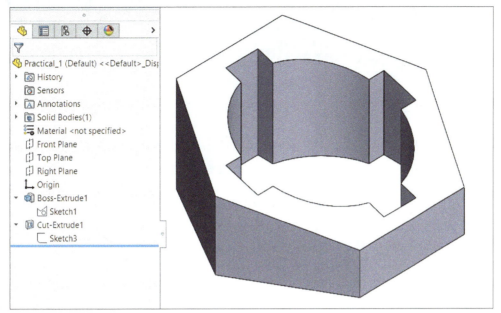

*Figure-2. Model for Practical 1*

- Right-click on sketch for first extrude feature and select the **Edit Sketch** button from shortcut menu toolbar; refer to Figure-3. The sketching environment will become active and sketch will be displayed; refer to Figure-4.

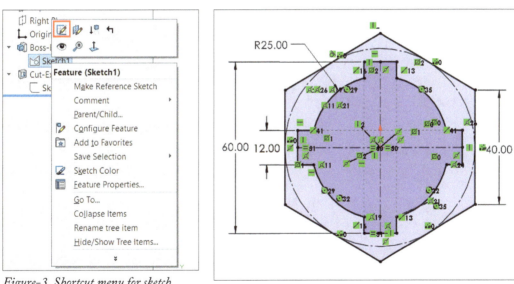

*Figure-3. Shortcut menu for sketch*

*Figure-4. Sketch*

- Note that the dimensions are applied without tolerances which is important manufacturing the part. To apply tolerance to a dimension, select it from the sketch. The **Dimension PropertyManager** will be displayed.
- Click in the **Tolerance Type** drop-down and select desired tolerance type; refer to Figure-5. We are using **Symmetric** option in our case.

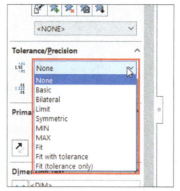

*Figure-5. Tolerance Type drop-down*

• Select the **Symmetric** option from the drop-down. The **Maximum Variation** edit box will be displayed below the drop-down.

• Specify desired tolerance value in the edit box; refer to Figure-6 and click on the **OK** button from **PropertyManager**. Note that for all distance dimensions, we will be applying 0.10 tolerance and for diameter/radius dimensions, we will be applying 0.05 tolerance.

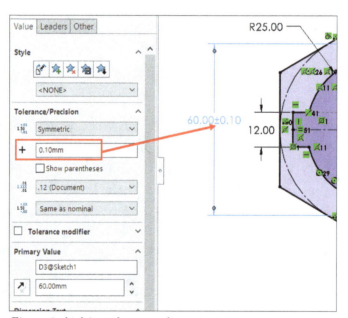

*Figure-6. Applying tolerance to dimension*

• Repeat these steps and apply tolerances to each of the dimensions in all sketches of model.

• Once you have applied tolerances and fully define all the sketches of model, click on the **Make Drawing from Part/Assembly** tool from **New** drop-down in **Quick Access Toolbar**; refer to Figure-7. The Drawing environment will become active and you will be asked to select desired sheet size from **Sheet Format/Size** dialog box; refer to Figure-8.

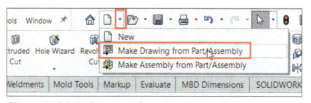

*Figure-7. Make Drawing from Part/Assembly tool*

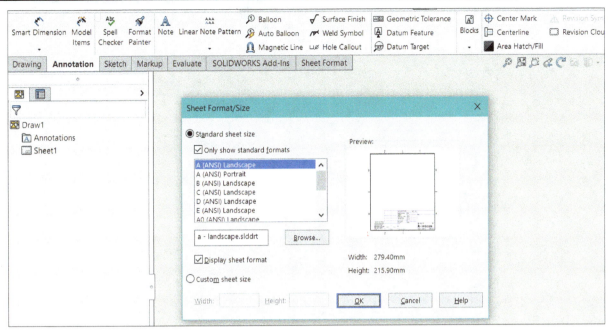

*Figure-8. Sheet Format Size dialog box*

- Generally, A4 size Landscape template is useful for printing these drawings because most of the workshops have small printers to print A4 size papers. Select the Landscape or Portrait depending on how long/wide your part is when placing views. Generally, landscape type suits most of the drawing views. Select the **A4 (ANSI) Landscape** option from list of templates and click on the **OK** button from the dialog box. The drawing page will be displayed with **View Palette** at right in the application window.
- Drag the Top view from **View Palette** and place it at desired location in the drawing page; refer to Figure-9. The **Projected View PropertyManager** will be displayed; refer to Figure-10 and projected view will be attached to the cursor.

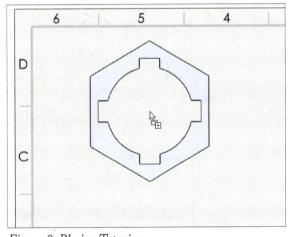

*Figure-9. Placing Top view*

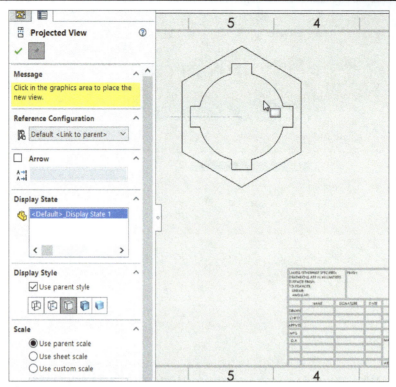

*Figure-10. Projected View PropertyManager*

• Click at desired location to place the projected view; refer to Figure-11. Press ESC to exit the tool.

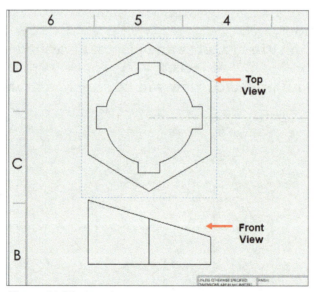

*Figure-11. Placing views*

## Importing Annotations in Views

• Click on the **Model Items** tool from the **Annotation CommandManager** in the **Ribbon**. The **Model Items PropertyManager** will be displayed; refer to Figure-12.

• Select the **Entire model** option from the **Source** drop-down in **Source/Destination** rollout of **PropertyManager** so that all the available annotations are imported in drawing views.

• Select the **Import items into all views** check box so that all the dimensions are imported in the all the views as possible. Note that software will not repeat the dimensions in various views which means if a dimension has been imported in Front view then it will not be displayed in Top view or Right view when importing dimensions.

*Figure-12. Model Items PropertyManager*

- Select desired buttons from **Dimensions**, **Annotations**, and **Reference Geometry** rollouts to import dimensions. Generally, we select the **Marked for Drawing**, **Instance/Revolution counts**, **Toleranced dimensions**, **Hole Wizard Profiles**, and **Hole Wizard Locations** buttons from **Dimensions** rollout and **Select all** check box from **Annotations** rollout to import dimensions.

- After setting desired parameters, click on the **OK** button from the **PropertyManager**. The dimensions will be applied to views in drawing; refer to Figure-13. Note that dimensions are not placed neatly.

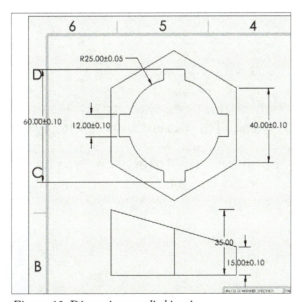

*Figure-13. Dimensions applied in views*

- Move the cursor to border of top view; refer to Figure-14 and drag the view when move cursor is displayed to middle of drawing sheet where all the dimensions can be accommodated inside the sheet. Note that all projected views will move accordingly.

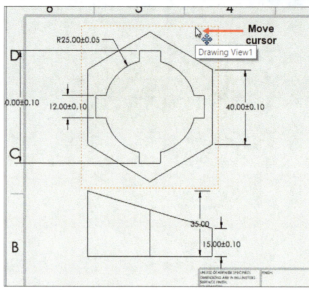

*Figure-14. Move cursor*

- Click on the dimension text which you want to move in drawing and drag it to the position where you want the dimension to be placed.
- After placing the dimensions properly, the drawing should display as shown in Figure-15.

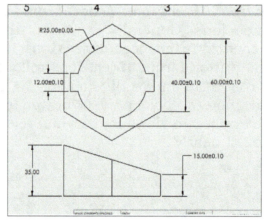

*Figure-15. After cleaning dimensions*

- Note that the 35 mm dimension is not having a tolerance applied to it. Select the 35 dimension from drawing. The **Dimension PropertyManager** will be displayed; refer to Figure-16.
- Select the **Symmetric** option from the **Tolerance Type** drop-down in the **Tolerance/ Precision** rollout and specify the tolerance value as **0.10**. Click on the **OK** button from **PropertyManager** to apply the changes.

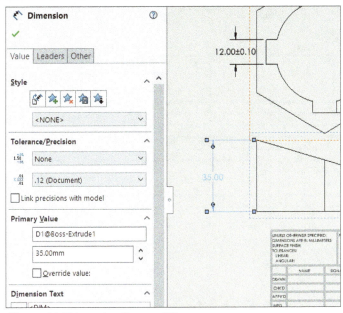

*Figure-16. Dimension PropertyManager*

## Editing Title Block

- Expand the **Sheet1** node from the **FeatureManager Design Tree** and right-click on **Sheet Format1** node. Shortcut menu will be displayed as shown in Figure-17.

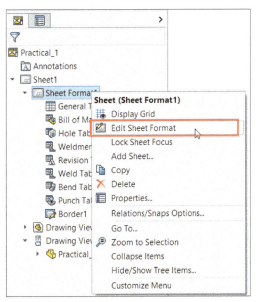

*Figure-17. Edit Sheet Format option*

- Select the **Edit Sheet Format** option from shortcut menu. The editing mode for sheet format will become active and input boxes of title block will become editable; refer to Figure-18.
- Double-click on any of the field to modify its parameter if note box is created already. To identify whether note box is already created in the field or not, hover the cursor inside the field. If the text edit cursor is displayed then note box has already be created in the field and you can modify it.

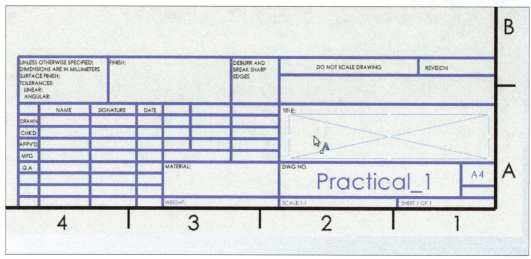

*Figure-18. Edit mode for Title block*

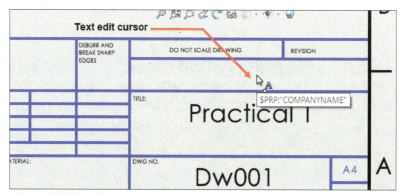

*Figure-19. Text edit cursor*

• After modifying title block, click on the exit button from top-right corner in the application window; refer to Figure-20.

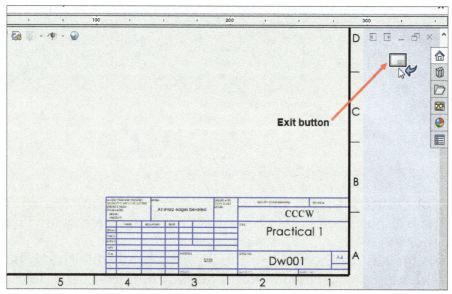

*Figure-20. Exit button*

• Click on the **Save** tool from **Quick Access Toolbar** or press **CTRL+S** to save the file. The **Save Modified Documents** dialog box will be displayed.
• Click on the **Save All** button from dialog box and save the files.

## PRACTICAL 2 (BASIC LEVEL)

Create drawing for the model shown in Figure-21.

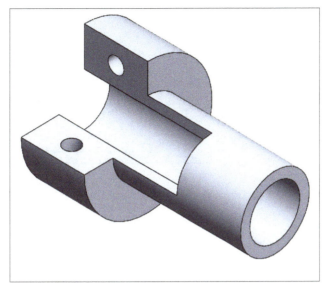

*Figure-21. Model for Drawing Practical 2*

Steps:
- Open the model in SolidWorks and make sure all the dimensions have been applied to the model with desired tolerances. This step ensures that the model is fully defined and you can directly import dimensions of model in drawing.
- Start a new drawing from Part Modeling environment and place the number of views that can fully define the model.
- Import the dimensions and apply more annotations if required.
- Define Title block parameters as desired and Save the file.

### Verifying Model Dimensions and Starting New Drawing Document
- Start SolidWorks if not started yet.
- Click on the **Open** tool from **Quick Access Toolbar** or press **CTRL+O**. The **Open** dialog box will be displayed.
- Select the **Practical 2** model file from resources folder of this chapter. The model will be displayed in graphics area.
- Modify the dimensions of sketches to include tolerances and make sure the sketches are fully defined so that all the dimensions to fully define model are imported in the drawing environment. Note that you can apply different precision for unit and tolerance as shown in Figure-22.

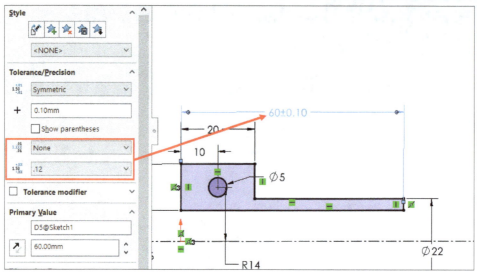

*Figure-22. Setting precision for unit and tolerance*

- There are two sketches used for creating model. After applying tolerances and fully defining the sketches, they should display as shown in Figure-23 and Figure-24. Make sure to save the file after making changes.

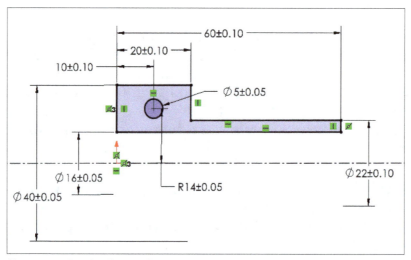

*Figure-23. Revolve feature sketch*

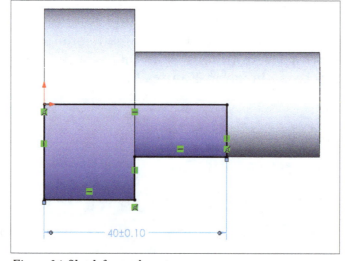

*Figure-24. Sketch for revolve cut*

- Click on the **Make Drawing from Part/Assembly** tool from the **New** drop-down in **Quick Access Toolbar**. The drawing environment will become active and **Sheet Format/Size** dialog box will be displayed.

- Select the **A4 (ANSI) Landscape** option from standard sheet size list and click on the **OK** button. The **View Palette** will be displayed at the right in application window and various views will be displayed in the palette.

## Placing Views and Applying Annotations

- Drag the Front view from **View Palette** and place it near the left corner of drawing sheet. The projected view will get attached to cursor; refer to Figure-25.

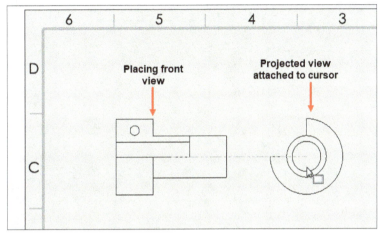

*Figure-25. Placing view*

- Click at the right of front view at desired distance to place the projected view. Press **ESC** to exit the view creation mode. After placing the views, you will notice that views are very small as compared to drawing sheet.

- To change the size of drawing views, select a view from the drawing sheet. The **Drawing View PropertyManager** will be displayed.

- Scroll down in the **PropertyManager** and select the **Use custom scale** radio button from the **Scale** rollout. The drop-down list below the radio button will become active; refer to Figure-26.

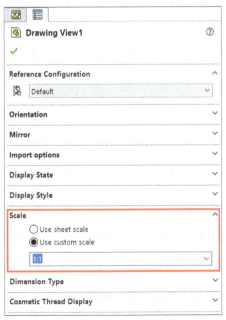

*Figure-26. Scale rollout*

- Selecting **2:1** from the drop-down list to double the size of views and select **1:2** from the drop-down list to make the size of views half. We are selecting **2:1** from the list. You can specify the scale in fractions like **1.5:1** by typing the value in edit box of this list and pressing **ENTER**. On selecting desired scale, size of the views will modify in the drawing sheet; refer to Figure-27.

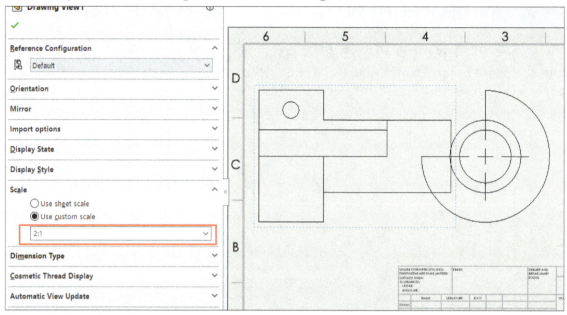

*Figure-27. On setting scale*

- Drag the views to desired locations to nearly place them at a proper distance from each other.
- Click on the **Model Items** tool from the **Annotation CommandManager** in the **Ribbon**. The **Model Items PropertyManager** will be displayed.
- Select the **Entire Model** option from **Source** drop-down, select the **Import items into all views** check box, and select desired buttons from various rollouts in the **PropertyManager**.
- After setting parameters, click on the **OK** button from the **PropertyManager**. The annotations will be created; refer to Figure-28. Note that some of the diameter dimensions should be placed in projected view rather than Front view.

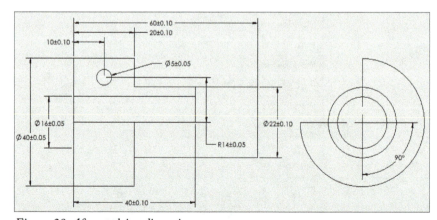

*Figure-28. After applying dimensions*

- Select the diameter dimensions and press **DELETE** button to remove the dimensions (dia 40, dia 16, and dia 22).
- Click on the **Smart Dimension** tool from the **Annotation CommandManager** in the **Ribbon**. The **Dimension PropertyManager** will be displayed; refer to Figure-29.

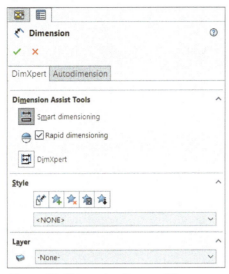

*Figure-29. Dimension PropertyManager*

- Select the outer edge of model in projected view. The radius dimension will get attached to cursor; refer to Figure-30.

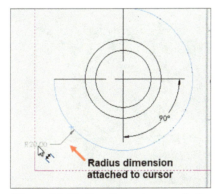

*Figure-30. Radius dimension attached to cursor*

- Click at desired location to place the dimension. Right-click on newly placed dimension and select the **Display As Diameter** option from shortcut menu; refer to Figure-31.

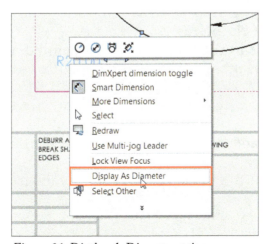

*Figure-31. Display As Diameter option*

- After changing radius dimension to diameter, select the dimension. The **Dimension PropertyManager** will be displayed. Set the tolerance as discussed earlier.
- Similarly, apply other diameter dimensions; refer to Figure-32.

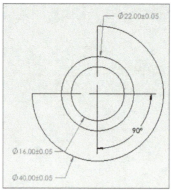

*Figure-32. After applying diameter dimensions*

- Set desired parameters in the Title block as discussed earlier and save the file.

## PRACTICAL 3 (MODERATE LEVEL)

Create drawing for the model as shown in Figure-33.

*Figure-33. Model for Drawing Practical 3*

### Starting New Drawing and Placing Views

- Start SolidWorks if not started yet and open the Practical 3 model file from the Resources folder of the chapter.
- Make sure the sketches in the model are fully defined and then click on the **Make Drawing from Part/Assembly** tool from the **New** drop-down in the **Quick Access Toolbar**.
- The Drawing environment will become active and the **Sheet Format/Size** dialog box will be displayed.
- Select the **A4 (ANSI) Landscape** option from the standard sheet size list and click on the **OK** button.
- Drag the Top view from the **View Palette** and place it near top left corner; refer to Figure-34.

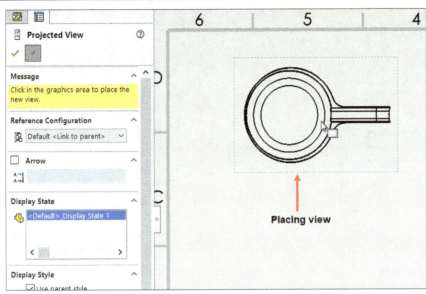

*Figure-34. Placing view*

- Press **ESC** to exit the tool. The Top view will be placed.
- Select the view from drawing sheet and set its scale value to **1:4** as discussed earlier.
- Click on the **Section View** tool from the **Drawing CommandManager** in the **Ribbon**. The **Section View Assist PropertyManager** will be displayed; refer to Figure-35.

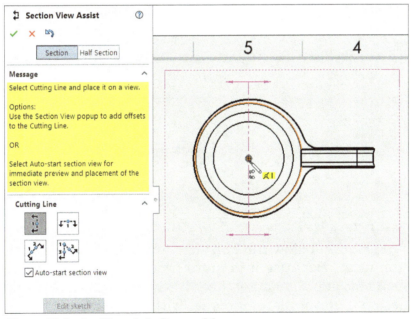

*Figure–35. Section View Assist PropertyManager*

- Click on the **Horizontal** button from the **Cutting Line** rollout of **PropertyManager**. The section line will be displayed as horizontal line; refer to Figure-36.

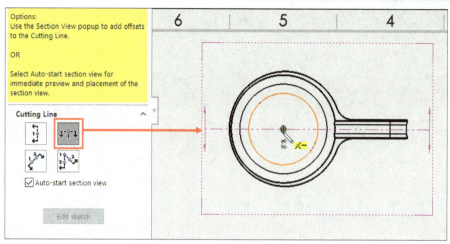

*Figure–36. Section line*

- Click at the center of circle. The section view will get attached to cursor; refer to Figure-37.

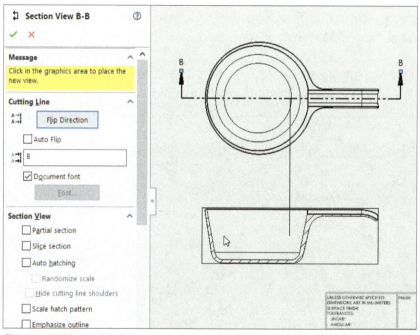

*Figure–37. Placing section view*

- Click at desired location below the Top view to place the section view. Why we are creating section view? Because it is more practical to show inner dimensions of model when you can actually show them in the view. As in this case, you cannot define thickness of bottom plate of model without using section view.
- Now, you need to create another section view for handle of model so that the thickness and dimensions of handle can be annotated in the view. Click on the **Section View Assist PropertyManager** will be displayed and section line will be attached to cursor.
- Make sure the **Vertical** button is selected in the **Cutting Line** rollout and click near the middle of handle; refer to Figure-38. The section view will get attached to the cursor.
- Place the view at the right of Top view at adequate distance for accommodating dimensions. Click on the **OK** button from the **PropertyManager** to create the view.

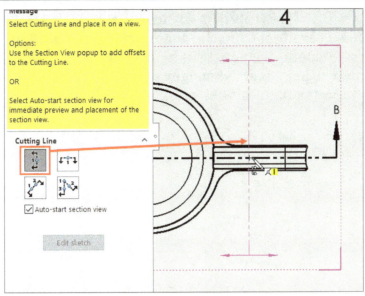

*Figure-38. Placing section line*

- Click on the **Projected View** tool from the **Drawing CommandManager** in the **Ribbon**. The **Projected View PropertyManager** will be displayed.
- Select the horizontal section view from the drawing sheet, the projected view will get attached to cursor; refer to Figure-39.

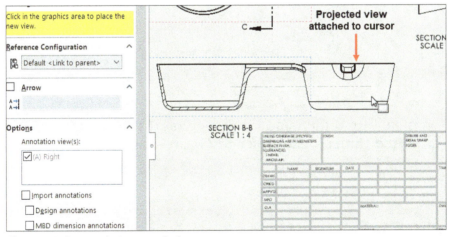

*Figure-39. Projected view attached to cursor*

- Click below the vertical section of the model to place this projected view; refer to Figure-40. Press **ESC** to exit the tool.

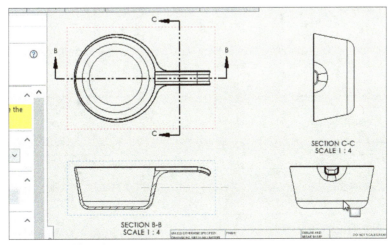

*Figure-40. After placing views*

## Applying Annotations

- Click on the **Model Items** tool from the **Annotation CommandManager** in the **Ribbon**. The **Model Items PropertyManager** will be displayed.
- Select the parameters in **PropertyManager** as shown in Figure-41 and click on the **OK** button. The dimensions of model will be applied in various views.

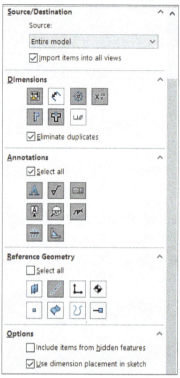

*Figure-41. Parameters in Model*
*Items PropertyManager*

- Separate the mixed up dimensions by dragging them one by one so that you can easily see imported dimensions; refer to Figure-42.

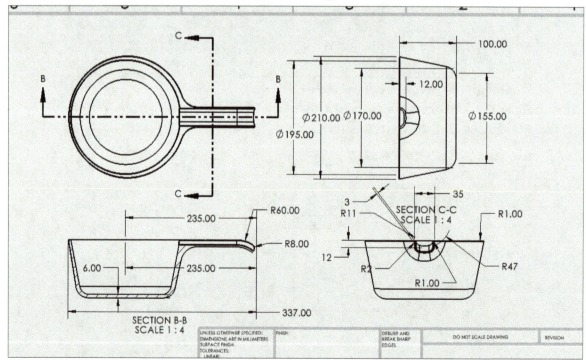

*Figure-42. After separating dimensions*

- Delete extra dimensions and apply the dimensions as shown in Figure-43. At some locations, you will need sketching tools of Drawing environment to create references for dimensioning. We have placed sketch centerlines at various locations in drawing views to apply dimensions.

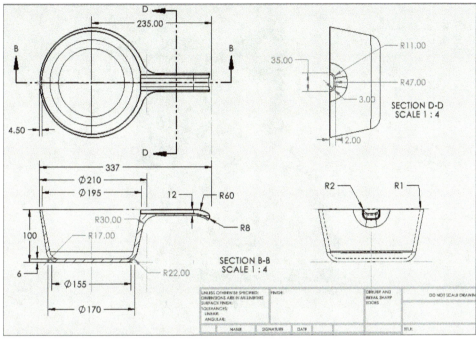

*Figure–43. After creating and managing annotations*

- Click on the **Note** tool from the **Annotation CommandManager** in the **Ribbon**. The **Note PropertyManager** will be displayed and a note box will be attached to cursor.
- Place the note box in empty area of drawing sheet and type the general notes for drawing like specifying radius for all fillets not dimensioned specifically; refer to Figure-44.

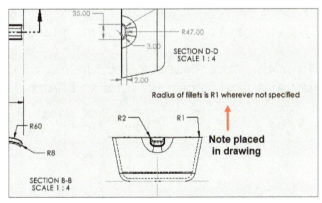

*Figure–44. Placing text note*

- Save the file and exit.

# PRACTICAL 4 (MODERATE LEVEL)

Create drawing for the model shown in Figure-45. Also, apply geometric tolerances as per the geometry of model.

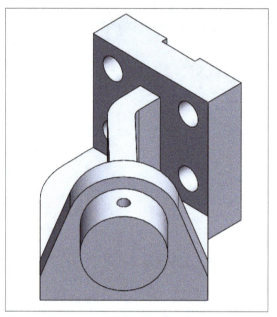

*Figure-45. Model for Drawing Practical 4*

## Starting New Drawing and Placing Views

- Start SolidWorks if not started yet and open the Practical 4 model file from the Resources folder of the chapter.
- Make sure the sketches in the model are fully defined and then click on the **Make Drawing from Part/Assembly** tool from the **New** drop-down in the **Quick Access Toolbar**.
- The Drawing environment will become active and the **Sheet Format/Size** dialog box will be displayed.
- Select the **A4 (ANSI) Landscape** option from the standard sheet size list and click on the **OK** button.
- Drag the Top view from the **View Palette** and place it near center of sheet. The projected view will get attached to cursor; refer to Figure-46.

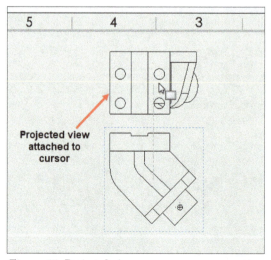

*Figure-46. Projected view*

- Click above the top view to place a projected view and press **ESC** to exit the tool.
- Note that the show dimensions of section model shown in Figure-47, you will need to use auxiliary view. For creating auxiliary views, you need to select an edge, axis, or sketch line to define direction for view. If you do not have a clear edge for creating then view then you can use sketch line tools.
- Click on the **Auxiliary View** tool from the **Drawing CommandManager** in the **Ribbon**. The **Auxiliary View PropertyManager** will be displayed and you will be asked to select an edge, axis or sketch line.

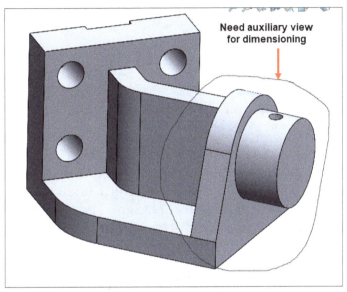

*Figure-47. Model section to be dimensioned*

- Click on the edge of model as shown in Figure-48. The auxiliary view will get attached to cursor.

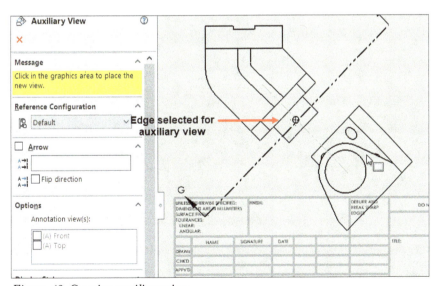

*Figure-48. Creating auxiliary view*

- Click at desired location in the drawing sheet to place the view; refer to Figure-49 and press **ESC**.

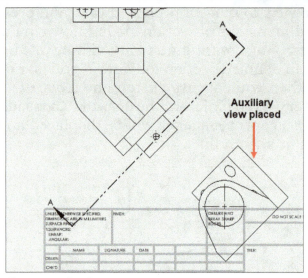

*Figure-49. Auxiliary view*

## Changing size of sheet

- Note that the space is not available to clearly place this view in the sheet. So, you need to change the size of sheet. To do so, right-click on **Sheet Format1** option from **Sheet1** node in **FeatureManager Design Tree** and select the **Properties** option from shortcut menu displayed; refer to Figure-50. The **Sheet Properties** dialog box will be displayed; refer to Figure-51.

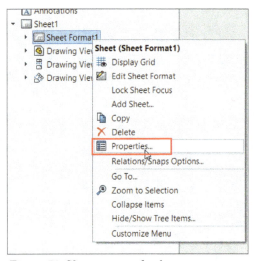

*Figure-50. Shortcut menu for sheet*

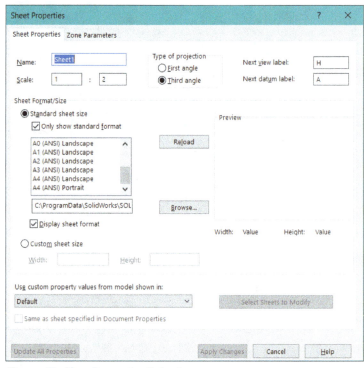

*Figure-51. Sheet Properties dialog box*

- Select the **A4 (ANSI) Portrait** option from template list in dialog box and click on the **Apply Changes** button. The sheet size will change to A4 portrait; refer to Figure-52. Drag the views to suitable location in the sheet; refer to Figure-53.

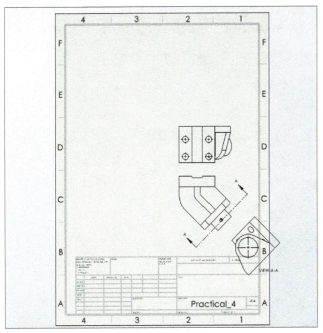

*Figure-52. A4 portrait sheet*

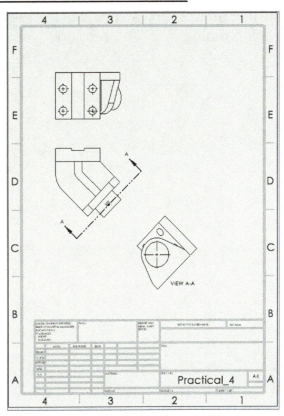

*Figure-53. Modified positions of views*

- In these three views, you can annotate most of the dimensions but you cannot annotate the height of rib feature unless you use hidden view which is generally kept as last resort for dimensioning. So, you need to create a section view that shows the height of rib feature. Click on the **Section View** tool from the **Drawing CommandManager** in the **Ribbon**. The **Section View PropertyManager** will be displayed.

- Select the **Vertical** button from **Cutting Line** rollout of **PropertyManager** and click on the mid point of top view; refer to Figure-54. The **Section View** dialog box will be displayed.

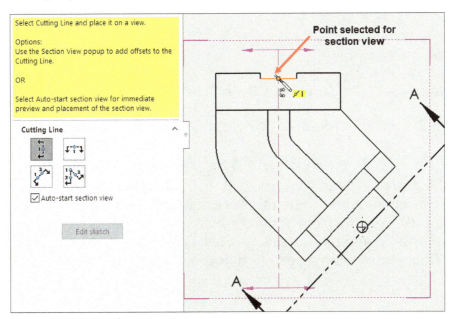

*Figure-54. Selection for section view*

- Click on the rib feature from the view; refer to Figure-55 and click on the **OK** button from dialog box. The section view will get attached to cursor; refer to Figure-56.

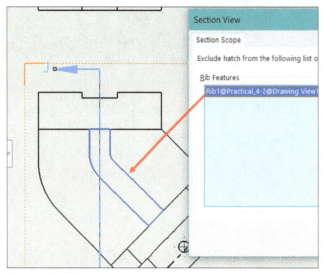

*Figure-55. Rib selected for section*

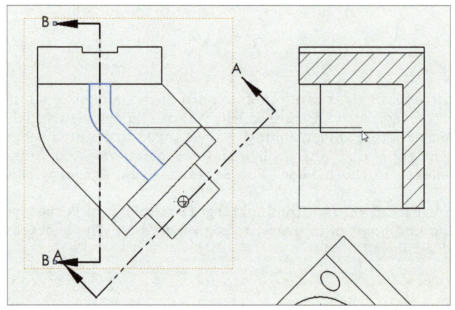

*Figure-56. Section view*

- Click at desired location to place the view and press **ESC** to exit the tool.

Now, you have all the views required to completely define the model with dimensions.

## Applying Annotations

- Click on the **Model Items** tool and set the parameters in **PropertyManager** as discussed earlier to define dimensions/annotations to be imported.
- Click on the **OK** button from **PropertyManager** to apply the annotations. The annotations will be applied in various views of drawing sheet; refer to Figure-57.
- Drag the dimensions to appropriate distances to make them clearly visible.

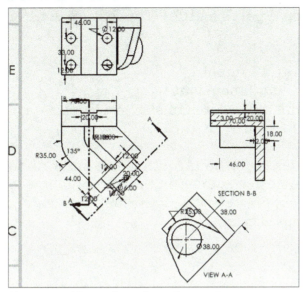

*Figure-57. After applying dimensions to views*

• Delete the unnecessary or wrong placed dimensions and apply the dimensions; refer to Figure-58. Make sure the model can be created based on the dimensions.

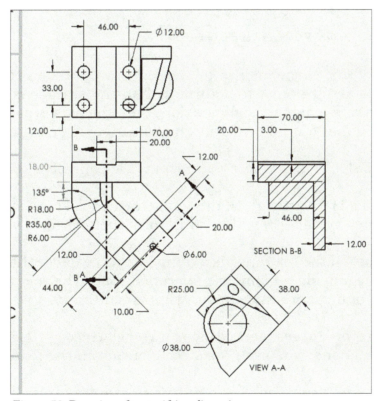

*Figure-58. Drawing after specifying dimensions*

## Applying Geometric Tolerances

Geometric Tolerances are used to define the parameters related to various features of model like flatness of a face, circularity of a cylinder, and so on. There are various factors which define the symbol type, tolerance value, and reference plane. A short note on geometric tolerance application is given next.

### Short Note on Geometric Tolerance

The feature control frame is also known as GD&T box in laymen's language. In GD&T, a feature control frame is required to describe the conditions and tolerances of a geometric control on a part's feature. The feature control frame consists of four pieces of information:

1. GD&T symbol or control symbol
2. Tolerance zone type and dimensions
3. Tolerance zone modifiers: features of size, projections...
4. Datum references (if required by the GD&T symbol)

This information provides everything you need to know about geometry of part like what geometrical tolerance needs to be on the part and how to measure or determine if the part is in specification; refer to Figure-59. The common elements of feature control frame are discussed next.

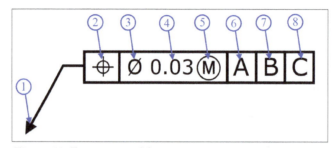

*Figure-59. Feature control frame*

1. **Leader Arrow** – This arrow points to the feature that the geometric control is placed on. If the arrow points to a surface than the surface is controlled by the GD&T. If it points to a diametric dimension, then the axis is controlled by GD&T. The arrow is optional but helps clarify the feature being controlled.

2. **Geometric Symbol** – This is where your geometric control is specified.

3. **Diameter Symbol (if required)** – If the geometric control is a diametrical tolerance then the diameter symbol (Ø) will be in front of the tolerance value.

4. **Tolerance Value** – If the tolerance is a diameter you will see the Ø symbol next to the dimension signifying a diametric tolerance zone. The tolerance of the GD&T is in same unit of measure that the drawing is written in.

5. **Feature of Size or Tolerance Modifiers (if required)** – This is where you call out max material condition or a projected tolerance in the feature control frame.

6. **Primary Datum (if required)** – If a datum is required, this is the main datum used for the GD&T control. The letter corresponds to a feature somewhere on the part which will be marked with the same letter. This is the datum that must be constrained first when measuring the part. Note: The order of the datum is important for measurement of the part. The primary datum is usually held in three places to fix 3 degrees of freedom

7. **Secondary Datum (if required)** – If a secondary datum is required, it will be to the right of the primary datum. This letter corresponds to a feature somewhere on the part which will be marked with the same letter. During measurement, this is the datum fixated after the primary datum.

8. **Tertiary Datum (if required)** – If a third datum is required, it will be to the right of the secondary datum. This letter corresponds to a feature somewhere on the part which will be marked with the same letter. During measurement, this is the datum fixated last.

### Reading Feature Control Frame

The feature control frame forms a kind of sentence when you read it. Below is how you would read the frame in order to describe the feature.

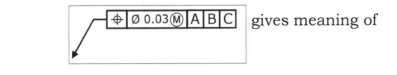 gives meaning of

Meaning of various geometric symbols are given in Figure-60.

| SYMBOL | CHARACTERISTICS | CATEGORY |
|--------|-----------------|----------|
| — | Straightness | Form |
| ▱ | Flatness | |
| ◯ | Circulatity | |
| ⌭ | Cylindricity | |
| ⌒ | Profile of a Line | Profile |
| ⌓ | Profile of Surface | |
| ∠ | Angularity | Orientation |
| ⊥ | Perpendicularity | |
| // | Parallelism | |
| ⊕ | Position | Location |
| ◎ | Concentricity | |
| ≡ | Symmetry | |
| ↗ | Circular Runout | Runout |
| ↗↗ | Total Runout | |

*Figure-60. Geometric Symbols*

Figure-61 and Figure-62 show the use of geometric tolerances in real-drawing.

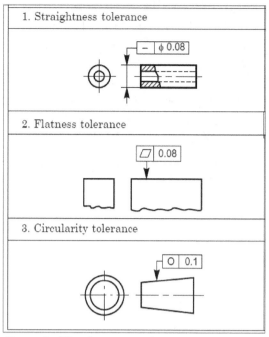

*Figure-61. Use of geometric tolerance 1*

Note that in applying most of the Geometrical tolerances, you need to define a datum plane like in Perpendicularity, Parallelism, and so on.

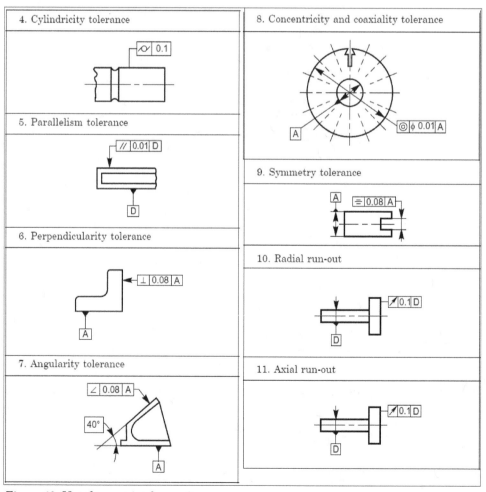

*Figure-62. Use of geometric tolerance 2*

There are a few dimensioning symbols also used in geometric dimensioning and tolerances, which are given in Figure-63.

| Symbol | Meaning | Symbol | Meaning |
|--------|---------|--------|---------|
| Ⓛ | LMC – Least Material Condition | ←⊕ | Dimension Origin |
| Ⓜ | MMC – Maximum Material Condition | ⊔ | Counterbore |
| Ⓣ | Tangent Plane | ∨ | Countersink |
| Ⓟ | Projected Tolerance Zone | ⊤ | Depth |
| Ⓕ | Free State | ⌀ | All Around |
| ⌀ | Diameter | ←→ | Between |
| R | Radius | ✕ | Target Point |
| SR | Spherical Radius | ▷ | Conical Taper |
| S⌀ | Spherical Diameter | ◁ | Slope |
| CR | Controlled Radius | □ | Square |
| ⓢⓣ | Statistical Tolerance | | |
| 77 | Basic Dimension | | |
| (77) | Reference Dimension | | |
| 5X | Places | | |

*Figure-63. Dimensioning symbols*

Based on the information given above, you need to specify geometric tolerances to faces which will be assembled to other parts. In this model, you need to specify flatness for the base face and flange wall as shown in Figure-64. Also, you need to specify cylindricity of the round boss feature on the model. We are assuming flatness to be 0.05 mm and cylindricity to be 0.01 mm. The steps are given next.

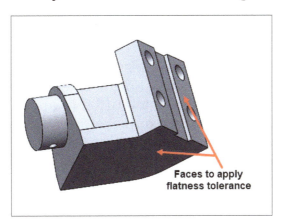

**Faces to apply flatness tolerance**

*Figure-64. Faces selected for flatness tolerance*

- Click on the **Geometric Tolerance** tool from the **Annotation CommandManager** in the **Ribbon**. The **Geometric Tolerance PropertyManager** will be displayed.
- Click on the face of model as shown in Figure-65. The symbol will get attached to cursor.
- Click at desired location to place the symbol. The options to define tolerance symbol will be displayed; refer to Figure-66.

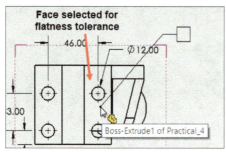

*Figure-65. Face to be selected*

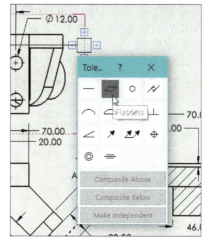

*Figure-66. Option to define tolerance symbol*

- Click on the **Flatness** button from dialog box. The **Tolerance** dialog box will be displayed; refer to Figure-67.

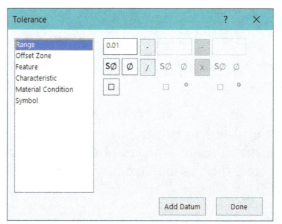

*Figure-67. Tolerance dialog box*

- Specify the value of tolerance in the **Tolerance** edit box of **Range** page in the dialog box. Since, there is no need to specify datum reference or other symbols for this geometric tolerance so click on the **Done** button to apply annotation. Click on the **OK** button from the **PropertyManager** to place the annotation.
- Click again on the **Geometric Tolerance** button and create flatness geometric tolerance for base face of model; refer to Figure-68.

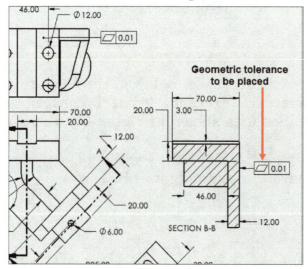

*Figure-68. Placing geometric tolerance*

- Click again on the **Geometric Tolerance** tool from the **Ribbon** and click on the diameter dimension of boss feature; refer to Figure-69. The **Tolerance** box will be displayed.

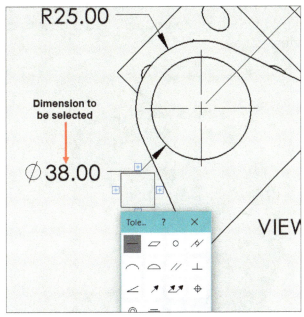

*Figure-69. Dimension selected for geometric tolerance*

- Click on the **Cylindricity** button from the dialog box. The **Tolerance** dialog box will be displayed.
- Click on the **Diameter** button from the dialog box to apply diameter symbol in tolerance value; refer to Figure-70.

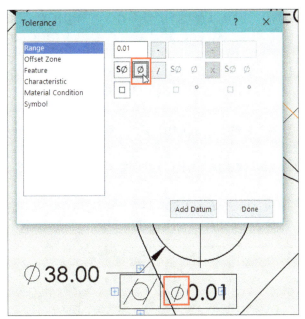

*Figure-70. Adding diameter symbol*

- Click on the **Done** button from the dialog box and then click on the **OK** button from the **PropertyManager**. Drag the annotation to desired location.
- Save the drawing file at desired location and close it.

# PRACTICAL 5 (ADVANCED LEVEL)

Create drawing for the model shown in Figure-71. This model is special because there are construction geometries in the model without which the model cannot be created. When you create drawing from the model, the construction geometries of sketches are not displayed by default so you will learn how to dimension construction geometries in various views.

*Figure-71. Model for Drawing Practical 5*

## Starting New Drawing and Placing Views

- Start SolidWorks if not started yet and open the Practical 5 model file from the Resources folder of the chapter.
- Make sure the sketches in the model are fully defined and then click on the **Make Drawing from Part/Assembly** tool from the **New** drop-down in the **Quick Access Toolbar**.
- The Drawing environment will become active and the **Sheet Format/Size** dialog box will be displayed.
- Select the **A4 (ANSI) Landscape** option from the standard sheet size list and click on the **OK** button.
- Drag the Top view from the **View Palette** and place it near center of sheet. The projected view will get attached to cursor; refer to Figure-72.

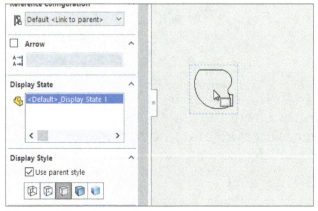

*Figure-72. Top view attached to cursor*

- Click at desired location to place the top view and other related views; refer to Figure-73. Press **ESC** to exit the tool.

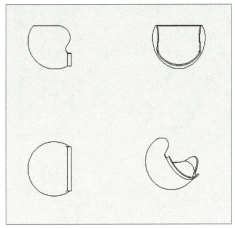

*Figure-73. After placing all views*

- Select the top view from drawing sheet and set the scale of view to 1:3; refer to Figure-74.

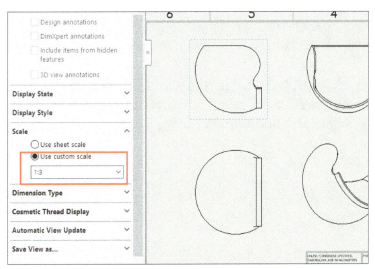

*Figure-74. Setting scale of view*

## Applying Annotations

- Click on the **Model Items** tool from the **Annotation CommandManager** in the **Ribbon**. The **Model Items PropertyManager** will be displayed.
- Set the parameters as shown in Figure-75 to import dimensions marked for drawing.

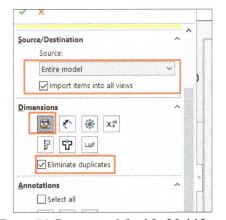

*Figure-75. Parameters defined for Model Items*

- Click on the **OK** button from **PropertyManager** to apply annotations. Drag the dimensions and views to make them neat; refer to Figure-76.

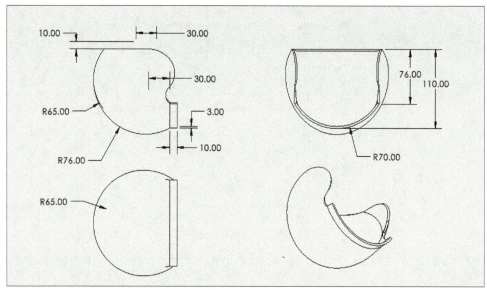

*Figure-76. After dragging dimensions and views*

- Note that the R65 dimension in projected view, distance 30 dimensions in top views are pointing to geometries which are not there in views.

## Making Geometries Visible in Views

- Expand the Drawing View2 > Practical 5 > Surface-Loft3 feature from **FeatureManager Design Tree** and select the **Show** option from Right-click shortcut menu of Sketch2 feature; refer to Figure-77. The sketch will become visible in view.

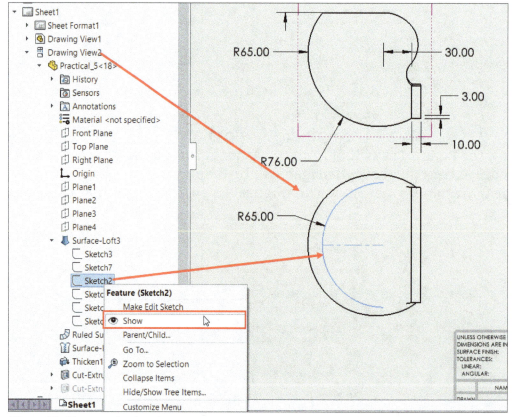

*Figure-77. Showing sketch in drawing view*

• Click on the **Centerline** tool from **Line** drop-down in **Sketch CommandManager** of the **Ribbon** and create centerlines to mark center points of various arcs in the views; refer to Figure-78. Note that dimension **30** and **10** at the top in Top view are showing geometry not displayed in the view. If you go through the 3D model and find feature with these dimensions then you will find that it is a cut feature not worth dimensioning as it is used to smoothen the face of model; refer to Figure-79.

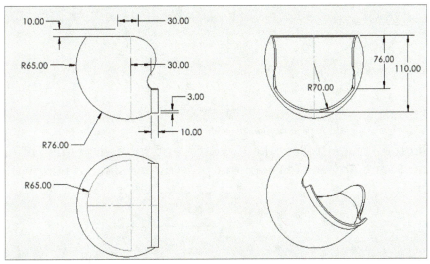

*Figure-78. After creating centerlines*

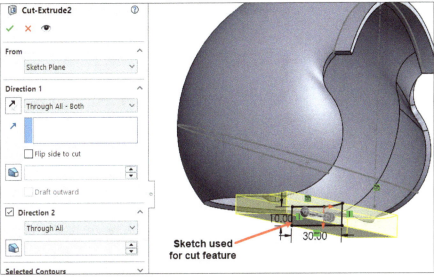

*Figure-79. Sketch used for cut feature*

• Select these dimensions from top view and delete them.
• Note that if you use **Smart Dimension** tool to dimension the arcs of top view then you will not be able to create dimensions; refer to Figure-80.

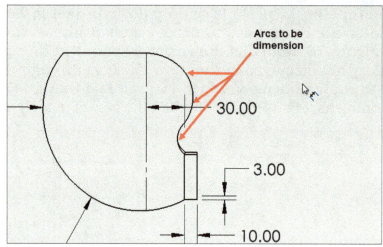

*Figure-80. Arcs to be dimensioned*

- Select the Sketch5 from Surface-Loft feature and make it visible; refer to Figure-81. Now, use the **Smart Dimension** tool to apply dimensions as shown in Figure-82.

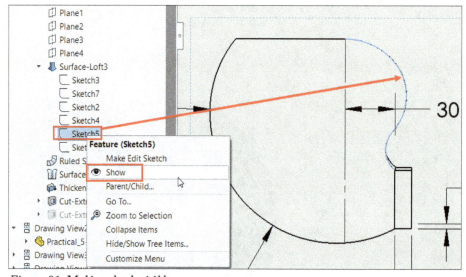

*Figure-81. Making sketch visible*

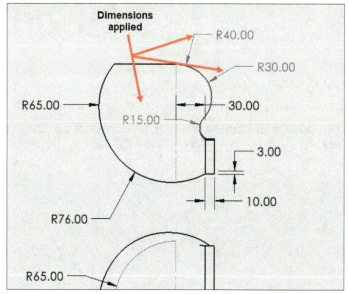

*Figure-82. Dimensions applied*

- Select the **Drawing View4** and make display style to **Shaded With Edges**; refer to Figure-83.

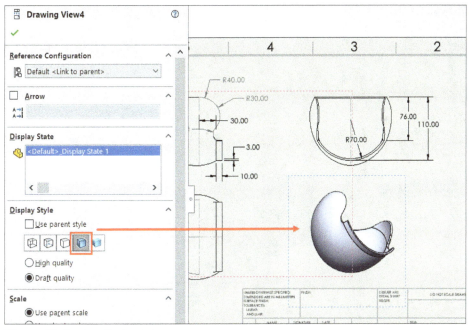

*Figure-83. Setting display style*

- Save the file at desired location and close it.

# PRACTICAL 6 (ASSEMBLY DRAWING)

Create drawing for the assembly as shown in Figure-84.

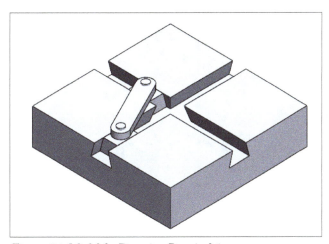

*Figure-84. Model for Drawing Practical 6*

Assembly drawings generally contain exploded views of assembly, bill of materials, and other related information to assemble the components. The steps to create assembly drawing for the model are given next.

## Exploding Assembly and Starting New Drawing

- Start SolidWorks if not started yet and open the Assembly.asm model file from **Practical 6** folder in the **Resources** folder of the chapter.
- Click on the **Exploded View** tool from the **Assembly CommandManager** in the **Ribbon**. The **Explode PropertyManager** will be displayed.

- Select all the parts from graphics area and set the parameters in **PropertyManager** as shown in Figure-85.

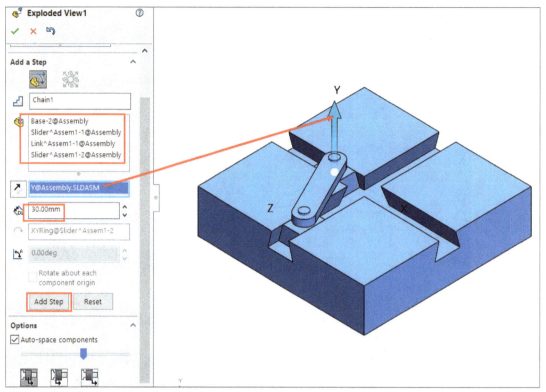

*Figure-85. Parameters for exploded view*

- Click on the **Add Step** button from the **PropertyManager** to create the exploded view; refer to Figure-86. Click on the **OK** button from **PropertyManager** to apply the changes. Press CTRL+S to save the file.

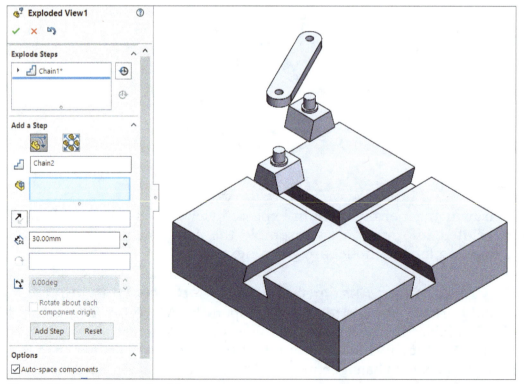

*Figure-86. Exploded view of model*

- Click on the **Make Drawing from Part/Assembly** tool from the **New** drop-down in the **Quick Access Toolbar**.
- The Drawing environment will become active and the **Sheet Format/Size** dialog box will be displayed.
- Select the **A4 (ANSI) Landscape** option from the standard sheet size list and click on the **OK** button.

## Creating Views and Applying Annotations

- Drag the Isometric Exploded view from the **View Palette** and place it near center of sheet; refer to Figure-87.

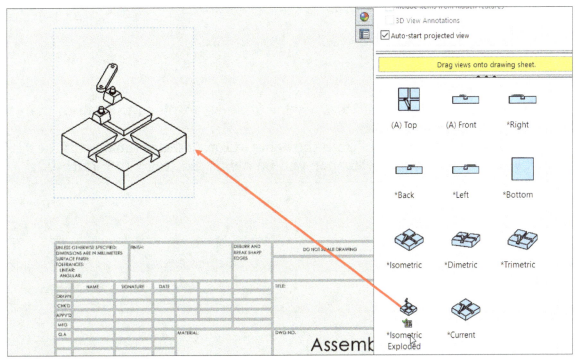

Figure-87. Isometric exploded view

- Select the newly placed view and set its scale to 1:3 to increase the size of model in view; refer to Figure-88.

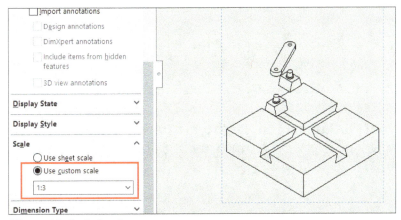

Figure-88. Scale for view

- Click on the **Bill of Materials** tool from the **Tables** drop-down in the **Annotation CommandManager** in the **Ribbon**. The **Bill of Materials PropertyManager** will be displayed and you will be asked to select drawing view for which you want to create bill of material.

- Click on the **OK** button from **PropertyManager**. The **BOM** table will get attached to cursor; refer to Figure-89.

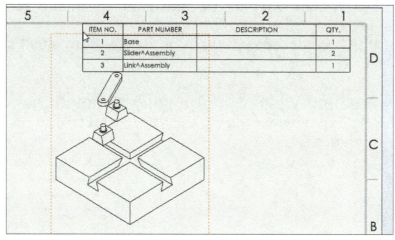

*Figure-89. Bill of materials*

- Click at desired location near top border of sheet to place the bill of materials. Drag the view to avoid overlap.
- Click on the **Auto Balloon** tool from the **Annotation CommandManager** in the **Ribbon**. The **Auto Balloon PropertyManager** will be displayed; refer to Figure-90.

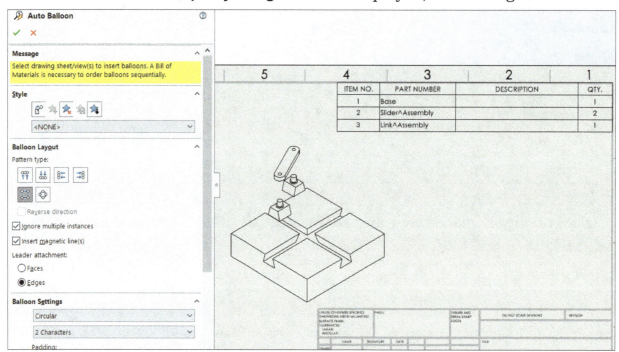

*Figure-90. Auto Balloon PropertyManager*

- Select the **Layout Balloons to Right** button to create balloons on the right of view and then select the view from drawing sheet. The balloons will be created; refer to Figure-91.

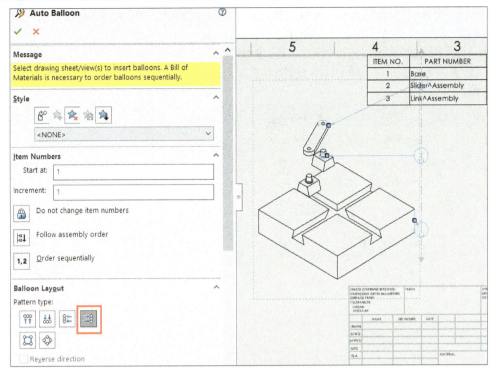

*Figure-91. Auto Balloon PropertyManager*

- Click on the **OK** button from **PropertyManager** to create balloons.
- Drag the magnetic line of balloons to change location of balloons; refer to Figure-92.

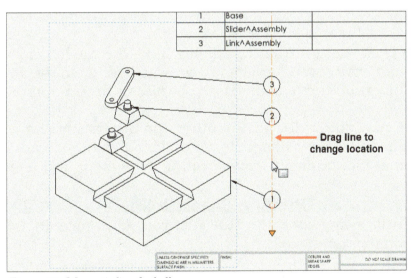

*Figure-92. Magnetic line for balloons*

## Adding Sheets and Part Drawings

- Click on the **Add Sheet** button from the bottom bar in the application window; refer to Figure-93. A new sheet will be added in the drawing and open automatically.

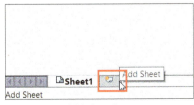

*Figure-93. Add sheet button*

- Click on the **Model View** tool from the **Drawing CommandManager** in the **Ribbon**. The **Model View PropertyManager** will be displayed.
- Click on the **Browse** button from the **PropertyManager** and select the part model file of your assembly to be inserted in the drawing sheet from the **Open** dialog box; refer to Figure-94.

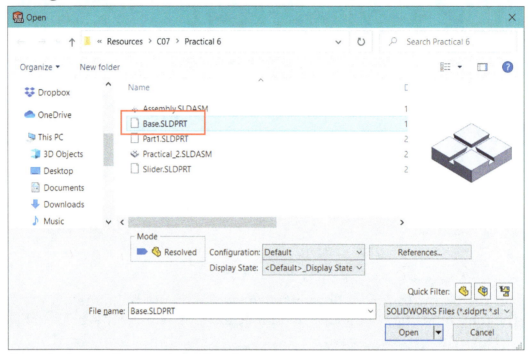

*Figure-94. Inserting Base part of assembly in drawing*

- Click on the Open button from the dialog box. The model will be imported for the drawing and front view of model will be attached to cursor. Place the views of model and dimension them as discussed earlier in this chapter.

Similarly, create new sheets for all the other components of the assembly. These steps are performed to make a complete set of drawings for current product so that the model drawings can be easily transferred to customers in pdf format.

## Printing Sheets to PDF

Note that to print drawing sheets to PDF file, you must have installed Adobe PDF reader application in your system.

- After creating drawing sheets for assembly and its parts, click on the **Print** tool from **File** menu. The **Print** dialog box will be displayed.
- Select **Adobe PDF** option (you can also use **Microsoft Print to PDF** option if available) from **Name** drop-down list at the top in the dialog box, select **All sheets** radio button from the **Print range** area, and set the number of copies to **1**; refer to Figure-95.
- Click on the **Preview** button if you want to check the preview of drawing sheets. Click on the **OK** button from the dialog box to create the file. The **Save PDF File As** dialog box will be displayed.
- Specify desired location and name of the file and then click on the **Save** button to save the file.

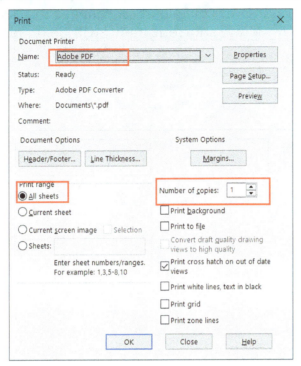

*Figure-95. Print dialog box*

• Save the file and close the document.

## PRACTICAL 7 (SHEETMETAL DRAWING)

Create sheetmetal drawing for the model shown in Figure-96. Drawings created for sheet metal parts generally include flat pattern of part and bend table with bends marked by balloons.

*Figure-96. Model for Drawing Practical 7*

### Creating Flat Pattern Configuration

• Start SolidWorks if not started yet and open the Practical 7 model file from the Resources folder of the chapter.

- Right-click on model name in **ConfigurationManager** at the left in the application window; refer to Figure-97. A shortcut menu will be displayed.

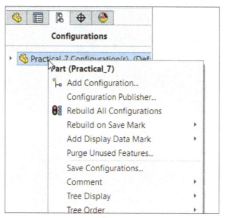

*Figure-97. Shortcut menu for configuration*

- Select the **Add Configuration** option from the shortcut menu. The **Add Configuration PropertyManager** will be displayed.
- Specify the name of configuration as Flat Pattern in **Configuration name** edit box and click on the **OK** button. The Flat Pattern configuration will become active.
- Right-click on roll back line in **FeatureManager Design Tree** and select the **Roll to End** option from the shortcut menu; refer to Figure-98. Alternatively, you can drag this line to the end of features to perform same task. On doing so, the **Flatten** tool will become active in the **Sheet Metal CommandManager** of the **Ribbon**; refer to Figure-99.

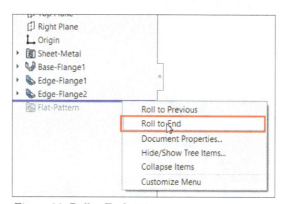

*Figure-98. Roll to End option*

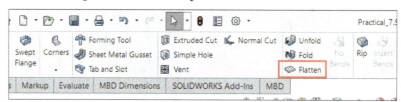

*Figure-99. Flatten tool*

- Click on the **Flatten** tool from the **Ribbon**. The model will be displayed as shown in Figure-100.
- Right-click on the **Default** configuration from the **ConfigurationManager** and select the **Show Configuration** option from the shortcut menu; refer to Figure-101. Setting this option will display model in original state.

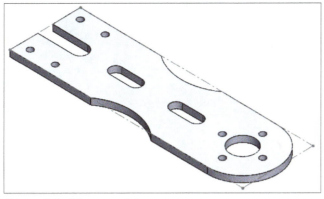

*Figure-100. Flatten model*

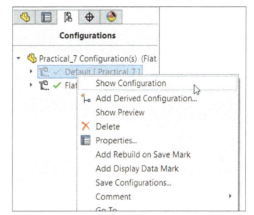

*Figure-101. Show Configuration option*

## Starting New Drawing and Creating Views

- Click on the **Make Drawing from Part/Assembly** tool from the **New** drop-down in the **Quick Access Toolbar**.
- The Drawing environment will become active and the **Sheet Format/Size** dialog box will be displayed.
- Select the **A4 (ANSI) Landscape** option from the standard sheet size list and click on the **OK** button.
- Drag the Top view from the **View Palette** and place it near center of sheet. The projected view will get attached to cursor. Place the projected views of model as shown in Figure-102.

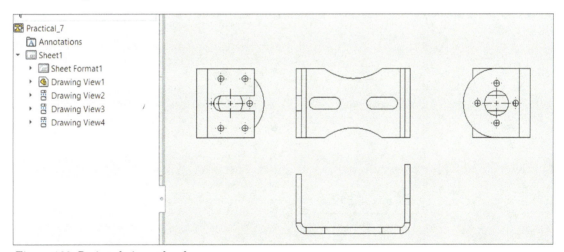

*Figure-102. Projected views placed*

- Select the Flat Pattern view from **View Palette** and place it in drawing sheet as shown in Figure-103. Press **ESC** to exit the tool.

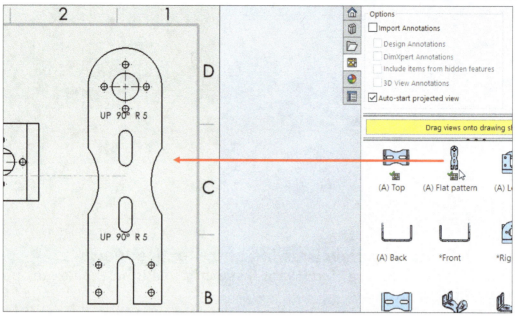

*Figure-103. Placing flat pattern view*

- By default, bend lines are not visible in flat pattern view. To display bend lines in drawing, select the **Bend Lines** option from **Show/Hide** cascading menu of the **View** menu; refer to Figure-104.

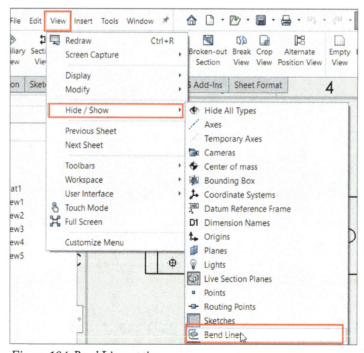

*Figure-104. Bend Lines option*

## Applying Annotations

- Click on the **Model Items** tool from the **Annotation CommandManager** in the **Ribbon**. The **Model Items PropertyManager** will be displayed.
- Set the parameters in **Model Items PropertyManager** as discussed earlier and click on the **OK** button. Rearrange the dimensions to place them appropriately.
- Click on the **Bend Table** tool from **Tables** drop-down in **Annotation CommandManager** of the **Ribbon** and select the Flat Pattern view from the drawing sheet. The options in **PropertyManager** will be displayed as shown in Figure-105.

*Figure-105. Bend Table PropertyManager*

- Select the **A, B, C** radio button or **1, 2, 3** radio button to define how bends will be marked in flat pattern and bend table.
- Click on the **OK** button from the **PropertyManager** to create the table. The table will get attached to cursor.
- Click at desired location to place the table; refer to Figure-106. Note that you can change the width of columns in table by dragging the separator line; refer to Figure-107.

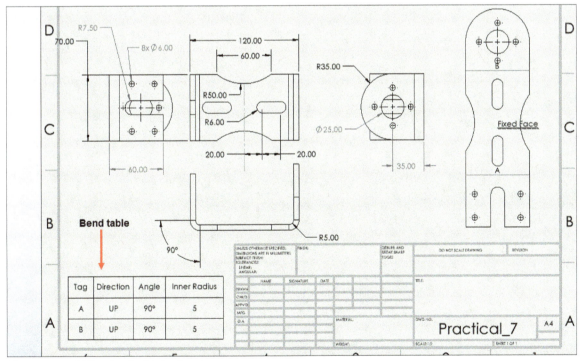

*Figure-106. Adding bend table in drawing*

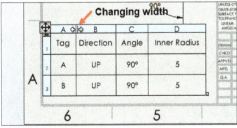

*Figure-107. Changing width of table*

- Specify the other parameters as discussed earlier and save the file at desired location.

## PRACTICE

Create drawing files for the models created under practice sections in Chapter 3, 4, 5, and 6.

## FOR STUDENT NOTES